CONNECTIONS

The EERI Oral History Series

Eric Elsesser

CONNECTIONS

The EERI Oral History Series

Eric Elsesser

Grace S. Kang, Christopher Arnold,
and Robert Reitherman, Interviewers

Earthquake Engineering Research Institute

Editor: Gail Hynes Shea, Berkeley, California, www.gailshea.com
Graphics: George Mattingly, Berkeley, California, www.mattinglydesign.com;
Book design: Laura Moger, Moorpark, California, www.lauramoger.com

The publication of this book was supported by FEMA/U.S. Department of Homeland Security under grant #EMW-2010-RC-51-K007.

Published by the Earthquake Engineering Research Institute
499 14th Street, Suite 320
Oakland, California 94612-1934
Tel: (510) 451-0905 Fax: (510) 451-5411
E-mail: eeri@eeri.org
Website: http://www.eeri.org

EERI Publication Number: OHS-18
Library of Congress Cataloging-in-Publication Data

Kang, Grace S.
Eric Elsesser / Grace S. Kang, Christopher Arnold, and Robert Reitherman, interviewers.
p. cm. — (Connections, the EERI oral history series ; 18)
Includes index.
ISBN 978-1-932884-47-0
1. Elsesser, Eric. 2. Civil engineers—California—Interviews. 3. Earthquake engineering—California—History. 4. Earthquake resistant design—History. I. Arnold, Christopher, 1925– II. Reitherman, Robert, 1950– III. Title.
TA140.E445K36 2010
624.092—dc22 2010051398

Printed in the United States of America
1 2 3 4 5 6 7 8 20 19 18 17 16 15 14 13 12 11 10 09

Acknowledgments

EERI gratefully acknowledges partial funding for production of this volume by the Elsesser family, Forell/Elsesser Engineers, David A. Friedman, Jim Guthrie, Grace S. Kang, Simin Naaseh, Paul Rodler, and Mason Walters.

Table of Contents

The EERI Oral History Series ix

Foreword. xiii

Personal Introduction by Christopher Arnold xv

Chapter 1 Growing Up in San Francisco 1

Chapter 2 Studying at Stanford 5

Chapter 3 Working for John Blume 9

Chapter 4 Starting His Own Practice 13

Chapter 5 Forell/Elsesser Engineers 17

Chapter 6 Engineering Projects 23

Chapter 7 Isolation and Other Seismic Protection Innovations. 33

Chapter 8 A Philosophy of Seismic Design 43

Chapter 9 Engineering Education and Research 49

Chapter 10 Thoughts on Architects and Engineers 55

Chapter 11 Collecting and Appreciating Art 59

Photographs and Illustrations. 65

Index 85

The EERI Oral History Series

This is the eighteenth volume in the Earthquake Engineering Research Institute's, *Connections: The EERI Oral History Series*. EERI began this series to preserve the recollections of some of those who have had pioneering careers in the field of earthquake engineering. Significant, even revolutionary, changes have occurred in earthquake engineering since individuals first began thinking in modern, scientific ways about how to protect construction and society from earthquakes. The *Connections* series helps document this important history.

Connections is a vehicle for transmitting the fascinating accounts of individuals who were present at the beginning of important developments in the field, documenting sometimes little-known facts about this history, and recording their impressions, judgments, and experiences from a personal standpoint. These reminiscences are themselves a vital contribution to our understanding of where our current state of knowledge came from and how the overall goal of reducing earthquake losses has been advanced. The Earthquake Engineering Research Institute, incorporated in 1948 as a nonprofit organization to provide an institutional base for the then-young field of earthquake engineering, is proud to help tell the story of the development of earthquake engineering through the *Connections* series. EERI has grown from a few dozen individuals in a field that lacked any significant research funding to an organization with about 2,500 members today. It is still devoted to its original goal of investigating the effects of destructive earthquakes and publishing the results through its reconnaissance report series. EERI brings researchers and practitioners together to exchange information at annual meetings and, via a now-extensive calendar of conferences and workshops, provides a forum through which individuals and organizations of various disciplinary backgrounds can work together for increased seismic safety.

The EERI oral history program was initiated by Stanley Scott (1921-2002). The first nine volumes were published during his lifetime, and manuscripts and interview transcripts he left to EERI are resulting in the publication of other volumes for which he is being posthumously credited. In addition, the Oral History Committee is including further interviewees within the program's scope, following the Committee's charge to include subjects who: 1) have made an outstanding career-long contribution to earthquake engineering, 2) have valuable first-person accounts to offer concerning the history of earthquake engineering, and 3) whose backgrounds, considering the series as a whole, appropriately span the various disciplines that are included in the field of earthquake engineering.

Scott's work, which he began in 1984, summed to hundreds of hours of taped interview sessions and thousands of pages of transcripts. Were it not for him, valuable facts and recollections would already have been lost.

Scott was a research political scientist at the Institute of Governmental Studies at the University of California at Berkeley. He was active in developing seismic safety policy for many years and was a member of the California Seismic Safety Commission from 1975 to 1993. Partly for that work, he received the Alfred E. Alquist Award from the Earthquake Safety Foundation in 1990.

Scott received assistance in formulating his oral history plans from Willa Baum, then director of the University of California at Berkeley Regional Oral History Office, a division of the Bancroft Library. An unfunded interview project on earthquake engineering and seismic safety was approved, and Scott was encouraged to proceed. Following his retirement from the University in 1989, Scott continued the oral history project. For a time, some expenses were paid from a small grant from the National Science Foundation, but Scott did most of the work pro bono. This work included not only the obvious effort of preparing for and conducting the interviews themselves, but also the more time-consuming tasks of reviewing transcripts and editing the manuscripts to flow smoothly.

The *Connections* oral history series presents a selection of senior individuals in earthquake engineering who were present at the beginning of the modern era of the field. The term "earthquake engineering" as used here has the same meaning as in the name of EERI—the broadly construed set of disciplines, including geosciences and social sciences as well as engineering itself, that together form a related body of knowledge and collection of individuals that revolve around the subject of earthquakes. The events described in the *Connections* series span many kinds of activities: research, design projects, public policy, broad social aspects, and education, as well as interesting personal aspects of the subjects' lives.

Published volumes in
Connections: The EERI Oral History Series

Henry J. Degenkolb 1994
John A. Blume 1994
Michael V. Pregnoff and John E. Rinne 1996
George W. Housner 1997
William W. Moore 1998
Robert E. Wallace 1999
Nicholas F. Forell 2000
Henry J. Brunnier and Charles De Maria 2001
Egor P. Popov 2001
Clarence R. Allen 2002
Joseph Penzien 2004
Robert Park and Thomas Paulay 2006
Clarkson W. Pinkham 2006
Joseph P. Nicoletti 2006
LeRoy Crandall 2008
Vitelmo V. Bertero 2009
Robert V. Whitman 2009
Eric Elsesser 2010

EERI Oral History Committee

Robert Reitherman, *Chair*
Thalia Anagnos
William Anderson
Roger Borcherdt
Gregg Brandow
Robert D. Hanson
Loring A. Wyllie, Jr.

Foreword

The interviews for this oral history occurred in California in the summer of 2007 at the Elsesser residence and at the Forell/Elsesser Engineers office in San Francisco. They were primarily conducted by Grace Kang, structural engineer and partner at Forell/Elsesser Engineers and a longtime colleague of Eric Elsesser's. Christopher Arnold and I participated in some of the interviews, and Eric's wife, Sylvia, assisted in providing additional information.

Grace and Eric had discussed the idea of documenting his "story" several times in the past. The interviews were begun by Grace when Eric received the stark news that he had been diagnosed with a brain tumor and had only months to live. His good humor and sharp intellect in the interviews that were conducted in that context come through clearly in these pages, and Grace, Chris, and I will certainly remember those occasions in a very personal way as well. Eric had time to review the manuscript and make corrections and additions prior to his passing, which occurred December 6, 2007.

The manuscript was reviewed by EERI Oral History Committee member Loring Wyllie and by Eric's wife, Sylvia. Further editing and indexing was accomplished by series editor Gail Shea, and page layout was by George Mattingly of GMD. Eloise Gilland, EERI Publications Manager, also assisted in the production of this book.

Robert Reitherman
Chair, EERI Oral History Committee
July 2010

Personal Introduction

Eric Elsesser's oral history reveals a brilliant student, a fruitful professional partnership, the evolution of a solid yet innovative engineering practice over several decades, the inspiration and training of a skilled staff, and the design of a succession within the office that would ensure its continued productive survival.

I met Eric in the mid-1960s, when we worked together on a research and development project. Building Systems Development (BSD), the architectural firm I had co-founded in 1964, was developing a school construction system that employed performance specifications as a bidding tool, and Eric was our structural consultant. Then in 1968, BSD began the Academic Building Systems (ABS) program for the University of California and Indiana University, for which Eric was, again, our consultant. As we worked together we realized, without particularly noting it, that we shared many common ideas about the relationships between structure and architecture. As Eric has noted in the pages that follow, the ABS program gave him the opportunity to develop and explore the characteristics of a huge array of structural systems—as opposed to the structural consultant's usual role of providing the architect with one system as quickly and cheaply as possible.

While we were working on the ABS project, the San Fernando earthquake of 1971 occurred. Eric would make occasional presentations to our staff after his reconnaissance visits to the area. A year or so after the earthquake, Eric suggested to me that we respond to a research solicitation by the U.S. Geological Survey with a proposal on architectural configuration. At the time, I had little idea of the full implications of architectural configuration on seismic performance. The USGS responded by telling us that the subject did not fall within its scope, and that we should send it to NSF. This gracious and helpful advice was our first bit of good fortune. The second was that Chuck Thiel, who was then heading up NSF's seismic division, responded very positively, saying that "I've been waiting for someone to send in a proposal on this subject." So we received a very large grant from NSF—it was one of the first grants that NSF awarded to an architectural proposal.

I had not realized that Eric already had a complete agenda for the subject, since he was already convinced of the importance of building configuration and had been trying to get it better recognized in the code. Bob Reitherman, at that time newly graduated from the architecture department at U.C. Berkeley—now interviewer and chair of this oral history series—and I worked on the study with Eric as our engineering consultant, and Eric made sure his ideas were properly presented. Our study appeared first as a report for NSF and was later published in 1982 as *Building Configuration and Seismic Design.**

Eric and I taught some graduate seminars together on building systems and seismic design for the Architecture Department at U.C. Berkeley in the late 1970s. In fact, Eric's input led me into a second career as a kind of facilitator between the architectural and engineering cultures—for which I have been eternally grateful. The conversations that Eric and I had enjoyed during the development of the configuration book continued for another three decades. I found that I was perhaps more interested in structure than most architects. In conversing with Eric, I tended to play the role of straight man, feeding him questions or bringing up topics for him to comment on. Eric would talk about—and draw—architectural/structural concepts in a way that few engineers were able to do. His conversation would typically progress through diagrams (often matrices), small, quick, lightning sketches of seismic systems, load/deformation curves, hysteresis loops and the like that together created a strange language not unlike Japanese pictographs.

In the late 1970s and early 1980s, supported at first by NSF and later by FEMA, Eric and I devised and conducted a series of two-day seminars around the country on seismic design for architects. Eric, myself, and Rich Eisner became a sort of traveling troupe and we developed quite a polished program and some good publications explaining seismic design for architects. In this series of seminars, about a dozen in number, I think Eric developed the style (twin screens) and content (hundreds of slides) that would culminate in the great series of lectures at Stanford when he was a visiting professor, in 2006, as part of the Bay Area's

* Christopher Arnold and Robert Reitherman, *Building Configuration and Seismic Design.* John Wiley and Sons, New York, 1982.

series of lectures on the centennial of the 1906 earthquake, and his Distinguished Lecture for EERI in 2007.

In 1984, Eric and I traveled to New Zealand together on an NSF grant to study seismic design of nonstructural components in New Zealand, where Eric talked with a number of innovative engineers and saw base isolation projects at firsthand. He made a number of contacts that, as he began to execute projects in the U.S., led to consulting work on base isolation in New Zealand.

In early 2001, I was approached by a U.K. publisher to write a seismic design book that might represent an update of *Building Configuration and Seismic Design*. I brought Eric in on the project and we secured a contract: we called the book *Architects Against Earthquakes*. However, in 2002, we were asked by FEMA, through EERI, to develop a publication that addressed architects, that would consist of chapters on the broad range of seismic design issues, each written by different authors (EERI members of course). The resulting FEMA publication is called *Designing for Earthquakes*.† Because of the FEMA project, and other constraints, we had to cancel the U.K. book, but not before we had enjoyed many productive planning sessions. Eric's chapter is one of the best expressions of his thinking that has ever appeared in print.

Eric was very persuasive. His partner, Nick Forell, once described him to me as "silver tongued." His style was quiet, relaxed, patient, and reasonable in tone, with an occasional raised eyebrow and aggrieved comment on something he thought foolish or incompetent. His relaxed manner belied an iron determination to push his ideas. I would often act as a sounding board for his periodic enthusiasms and discoveries. For a time, in the 70s he became intrigued with fabric roof structures and tried to persuade me of their virtues (but I have never liked them).

Like many perfectionists, Eric's ideas never stopped, and this made him difficult to work with on any project that had a deadline. When EERI asked Eric and myself to provide a slide

† *Designing for Earthquakes: A Manual for Architects*, FEMA 454. Federal Emergency Management Agency, Washington, DC, 2006.

show for the 1998 Annual Meeting that celebrated EERI's fiftieth anniversary, we had several months to design a lunchtime historical presentation on the role of EERI in the history of seismic design. We devised a complex twin screen slide carousel presentation that needed a very tight script and several days of rehearsal. The morning of the show, Eric was still rearranging his slides. We went on, "live," with no rehearsal, and the presentation was flawless. To work with Eric was to live on the edge and enjoy a particularly stressful kind of fun.

Chris Arnold
Palo Alto, California
July 2010

CONNECTIONS

The EERI Oral History Series

Eric Elsesser

Chapter 1

Growing Up in San Francisco

When I was ten years old, I decided I was going to be a structural engineer.

Elsesser: I was born May 2, 1933.

Since about fifth or sixth grade I was designing as an engineer. I would make models, send them into architectural contests—that sort of thing. When I was ten years old, I decided I was going to be a structural engineer.

Kang: What made you decide that?

Elsesser's Family

Elsesser: My father, John Elsesser, was a contractor and cabinetmaker. He would visit architects, and I would accompany him and wander around the office while he was meeting with the architect. I was interested in buildings, but I saw people drafting lots of little details and that didn't appeal to me. I wanted to design something big.

I also liked to wander around downtown San Francisco to see the large buildings going up. From the time I was eight, when I was in Sherman Elementary School in San Francisco, I was taking the Muni around the city. You could get on most any streetcar and it would take you down to Market Street. Construction was still occurring then because the U.S. wasn't yet in World War II. The war basically stopped any significant nondefense construction work. I liked seeing the big frameworks go up. I remember seeing the garage under Union Square being constructed.

Kang: You were more interested in the bones of the structure than the architectural details?

Elsesser: Yes. I was interested in the overall architectural effect of a completed building, but not the architectural details that take up so much of an architect's time. But I maintained a lifelong interest in architecture. In junior high, at Marina Junior High, I did a lot of artwork and enjoyed that. I kept that interest up in high school, at George Washington High School. It was a 45-minute bus ride across town to get to high school, and I usually had to be there by 7:30 for athletic practice. I did track, cross country, and soccer. I enjoyed the usual variety of liberal arts. I did well in the technical subjects and math, but I was always interested in all the broad scope of learning outside that area also.

Art and architecture and engineering have been interests of mine for a long time, and to me they are very closely related. We can talk later about the collection my wife Sylvia and I have.

My father was a free spirit and more—a real individual. He climbed solo many of the tallest mountains in the American West. I've also enjoyed the outdoors, and I've hiked around Mt. Rainier with my son, Adam, and also in the Wind River Range in Wyoming—great places. For the last forty years of his life (and he lived until he was 94), my father never ate cooked food. He said it killed the food.

He was in the Merchant Marine in the First World War, came home to New York City, and visited his mother in Hell's Kitchen where they lived after World War I. He said it was nice to see his mother, but he left for California.

Kang: He was born in New York?

Elsesser: Yes.

Kang: And his parents?

Elsesser: They came from Germany in 1896. He was born almost two years later, on October 25, 1897. He had a wild past. How did he learn to swim? His friends threw him off a bridge in New York City and he got to shore.

My mother, Rikee Richterman, when she was a young woman and before she met my father, hitchhiked from New York to California with some girlfriends, taking five months. In 1939, she took me to New York on the train, taking five days. In 1967 Sylvia and I took my mother back to New York when the Montreal World's Fair was going on. But once she had left New York, she ended up in California for good, and that's how she met my father.

My father divorced my mother when I was six and remarried within a few months. His second marriage lasted six months. He was unmarried for ten years, and then he married Trude (pronounced Trudeh) Guermonprez in 1951. She was one of the world's great weavers. She was born in Europe, married there, was in the underground in Holland, and her husband and other resistance members were shot the day before the Germans left at the end of World War II. Her mother and father were in this country during World War II, he a pianist, she a bookbinder. The whole family was creative.

My stepmother, Trude, taught at Black Mountain College.[1] She replaced Anni Albers, another world famous weaver. The Albers couple who had taught there, Anni and Josef, like many of the other faculty, had come from the Bauhaus and were refugees from Germany when Hitler took over in the 1930s.

Trude came to Pond Farm[2] up north on the California coast. That's where she and my father met. It was through my stepmother's strong connection with the arts and artists that I later ended up with quite an art collection. We had an original Paul Klee painting, and one by Gino Severini. My father had these pieces but didn't keep track of the art very well as he got older. The Klee got lost for a while and was found in a closet in my father's house.

1 Black Mountain College existed from 1933 to 1956 in North Carolina. It was a small, innovative college for art and the performing arts that included as instructors Willem de Kooning, Buckminster Fuller, Franz Kline, Walter Gropius, Robert Motherwell, and Ilya Bolotowsky, as well as Anni and Joseph Albers. It functioned as an egalitarian community with farm work and other tasks shared by faculty and students. The *Black Mountain Review* was an influential poetry journal.

2 Marquerite Wildenhain, a potter trained at the Bauhaus, established Pond Farm, a summer art school near Guerneville, California, in 1949 and operated it until 1980.

Beginnings of a Visual Approach to Engineering

Kang: When I told a former Forell/Elsesser engineer we were doing these interviews, he said, "How can you make a verbal record of what Eric says? When he talks, he draws just as much as he talks!" Your interest in art and design seems related to your stepmother's involvement in art and your father's craft with woodworking. Would you say that the strong visual sense you had—as early as elementary school—influenced the way you later approached structural design?

Elsesser: Yes, designing with the visual imagination is very much a part of the engineering I have done. I can't imagine it otherwise.

Chapter 2

Studying at Stanford

Jack Benjamin would say, "Design me a bridge—by tomorrow." It was total immersion in engineering.

Elsesser: I applied to two schools, Stanford and Berkeley. I received the Stanford letter first. It said welcome to Stanford—and we'll give you a full-tuition scholarship. The next day the Berkeley letter came, and I was offered no scholarship. So, it was an easy decision.

I started at Stanford in 1951. It was a totally different school then. The university consisted mostly of the Quad, Hoover Tower, the library, and a few small outlying buildings. You could park anywhere you wanted—because very few people had a car.

With all the returning veterans from the war, there was a shortage of on-campus housing. Some of the sophomores had to live in the Stanford Hospital in Menlo Park that had been put up during the war. The walls were a single layer of 1/4-inch plywood that you could just about shove your hand through.

Kang: What were your courses like?

Elsesser: I enrolled in the engineering degree program, and that set almost all your courses. I took the standard 15-unit set of courses to start with, but wanted to be able to study some additional subjects. Then I took 18 units, and found I had enough time to do more, so I took 21 units. I took an art class that motivated me to stay up late, designing mobile models that filled my dormitory room. My nose was just above the water level at that point, but I told myself that that was what I was there for—studying, going to all the lectures and foreign films on campus, and so on.

The second year, while Stanford still gave me a scholarship for my tuition and room, they economized by eliminating my board. So I joined an eating club, and instead of paying for food, I got my meals in exchange for washing dishes in the kitchen. For the next three years I did that for each meal. Lunch was quite a busy time—get across campus, eat, wash the dishes and get back to the next class in time. On Sundays, the eating clubs closed, so you had to search for your own food.

Kang: How did you manage on Sundays?

Elsesser: You can hear my wife, Sylvia, laughing in the next room—she was my meal ticket on Sundays.

Kang: When did you and Sylvia meet?

Elsesser: We knew each other slightly in high school, but it was really when we were freshmen at Stanford that we got to know each other. We got married four years later, when we received our baccalaureate degrees, then we both went on for a graduate degree, she in education. I did the master's program in structural engineering.

My undergraduate engineering classes at Stanford were very good, but in a way they were not nearly as interesting for me as the graduate program. I would visit the engineering library and go through lots of books that showed the new work being done in Europe, and that was an inspiration for me.

Master's Program and Jack Benjamin

Elsesser: Graduate school was totally different. You started each day with a four-hour course. It was rigorous. Jack Benjamin would say, "Design me a bridge—by tomorrow." It was total immersion in engineering.

Kang: What was Jack Benjamin like?

Elsesser: He was fantastic, if you really wanted to learn engineering. If anyone suggested the work load was too heavy, that you had other courses to tend to, he would say, "But I thought you wanted to be an engineer?" Not everyone made it through graduate school. It was what I was looking for—this was real engineering. I would work till one or two in the morning the whole damn year. It was exhausting. It was wonderful.

Kang: So it was a real application of what you had learned?

Elsesser: And if you hadn't learned it yet, you had to quickly figure it out. One time we walked into class and Professor Benjamin put a blue book on each student's desk. He wrote a question on the board no one knew how to answer and left the room. He returned a little while later, picked up our blank blue books, and said, "Okay, this is what we're going to learn today."

Kang: Explain a little bit about what design techniques you learned from Jack Benjamin that relate to seismic design.

Elsesser: You had to figure out where the points of contraflexure would occur in beams and columns and thereby determine how the base shear was distributed. The analysis of structures was based on deflected shapes. With the deflected shapes known, the structural analysis became simple.

Kang: When you were a student in the 1950s, did you have any women in your engineering class?

Elsesser: No, I don't think there were any, out of about forty graduate students.[3] Then after that year, I decided not to go on into the PhD program. I didn't take the test for it, but was recommended for it anyway and received a letter saying I could be admitted. I said no thanks, I wanted to get out and do design work.

3 Ruth Gordon, the first woman to achieve the California registration of Structural Engineer in 1959, graduated from Stanford a little earlier than Elsesser, getting a bachelor's degree in civil engineering in 1948 and master's in structures in 1949. (Stanford School of Engineering "Alumni profile, Ruth Gordon: Alumna brushed aside barriers to build legacy as structural engineer, role model," *Engineering Today*, Stanford University School of Engineering, summer 2008, 4-5). She was one of two women in her class gaining the undergraduate civil engineering degree and the only one in her class the next year to obtain the master's in structures. (SEAONC, "Ruth Gordon," http://06earthquake.org/ruth-gordon, 2006).

Chapter 3

Working for John Blume

If you have a good grasp of design, it makes the analysis go smoothly.

Elsesser: My first engineering job out of college was with John Blume, in 1956. When Sylvia and I finished our master's degrees, we didn't go to our own graduation ceremony from Stanford. Instead, we went to the Sierra for a few days of rest, came back, and I went right to work. No trip abroad or any of those luxuries, because we had no money. We found a nice little flat to rent in San Francisco on Russian Hill for forty dollars a month. I worked for John Blume for a little over four years, and during that time I got my California structural engineer license.

Kang: Do you recall what task Blume put you to work on when you were hired?

Seismic Design of Tall Buildings

Elsesser: My first project at the Blume office was the city administration building for Auckland, New Zealand. It was the first construction in New Zealand to apply the latest in earthquake engineering to a tall building.[4] I was told we were the consultants for the

4 When construction of the Civic Building was completed in 1966, it was the tallest building in New Zealand, twenty stories and 233 ft (71 m) in height.

seismic design. There was some early analog computer analysis done and the preliminary design I worked on checked out very well.

Then John got a joint consulting job for the Bethlehem Steel Building in San Francisco, and asked me to do the preliminary sizes of structural members for the whole building. So I did that preliminary design and then John said we had to do a moment distribution analysis. I said it would take two months to do that—today such calculations are trivial with current computers and software. So I did that and it checked out. If you have a good grasp of design, it makes the analysis go smoothly. I picked up that knack at Stanford. You select the points of contraflexure in beams and columns and quickly get to answers that were within 10 percent of a more exact, laborious approach.

We also had the job of quickly analyzing a tall hotel in Portland, Oregon as it was about to be constructed, in 1957. It had big columns and small beams. I calculated the moment capacity of the beams and columns. I started to draw the diagram of the moment capacity story by story, and concluded that the slabs did not have the capacity to force rotations into the columns until the twentieth story was reached. They were using the wrong analysis. I went into John's office and said, "I'm probably completely wrong, but this structure doesn't work." John had Rol Sharpe and some other engineers come and talk about it, and we agreed it made no sense. So we went to Skidmore, Owings & Merrill (SOM) , told them about the problem, and the structure was re-designed. There was a big cost overrun.

Steve Johnston, chief structural engineer with SOM, thought about the dynamic response of tall buildings, and concluded the usual code approach and standard analysis was not really telling you how the building would actually perform.

At this time, I was making little wire structural models of designs that were going on in the Blume office. Our landlord at home where we were renting let me have access to an unused area in the basement for that.

SEAOC Blue Book

Kang: Were you involved in the development of the Blue Book in the 1950s?[5]

Elsesser: I was on the statewide SEAOC Seismology Committee when I worked for Blume. There were only six of us: two from San Francisco, two from Los Angeles, one from San Diego, one from Sacramento. After I opened my office, I continued with the committee work for a few years and then decided I would take about five years off from that and devote my time to developing my practice. I was on the committee in the late fifties and early sixties, then took time off the latter half of the sixties, and then went back on the committee up through the seventies. I was especially interested in provisions for irregular buildings, and finally in the mid-1970s got something adopted. But the solution adopted was to require more analysis, rather than to

5 Seismology Committee, Structural Engineers Association of California, *Recommended Lateral Force Requirements and Commentary*, first edition 1959.

prohibit the serious irregularities. That wasn't my approach. After that, I devoted the rest of my life to designing good buildings, the best structures I could create, rather than spend my time worrying about code provisions that allowed for minimal buildings.

Leaving the Blume Firm

Elsesser: While I was working for Blume, I was making $2,500 a year, and I figured I could make a better living on my own. I was getting good-paying work on my own, without really pushing much. Sylvia and I managed financially with a combination of my regular income, the extra money from my own work, and her income from teaching at a public school in San Francisco. Our first child, Linnea, was born in December of 1959, followed by our son Adam in 1961. Both of their families now live in the San Francisco area. Adam has four children, and Linnea now has two daughters. Back in the mid-1950s, with two children, it was obvious I should strike out on my own and make a better living.

This was when Blume had just gotten the big SLAC job, the structure of the Stanford Linear Accelerator. I went into John's office [1960] and said, "John, you're not going to like what I'm going to tell you. I'm leaving to start my own practice." He said I would be missing some big, exciting work and I said I knew that, but I wanted to go out on my own. I asked him for a favor. If I bombed out in my new venture, could I come back to work for him? He said yes, but I would have to start over at the bottom rung. Fortunately, that never came to pass.

Chapter 4

Starting His Own Practice

Those were the days when the engineer could walk on the site, tell the contractor what he had done incorrectly, and then the contractor would thank you for the correction and get it done right.

Kang: What was one of the first projects you did on your own?

Elsesser: The architecture firm of Marquis and Stoller, the San Francisco architecture firm, had the job of designing the large housing project in the Western Addition—St. Francis Square, a three-block multi-family housing project—for the International Longshore and Warehouse Union. It was built for about $9 per square foot.

Projects for Marquis and Stoller

Elsesser: I recall Bob Marquis coming over to our house when we lived on Hyde Street, where we displayed and enjoyed our art-work. Having an aesthetic sense helped me develop a relationship with Marquis and his partner Claude Stoller. They offered to give me some other jobs if I opened my own office. At that time, I had my civil, but not yet my structural registration, which I obtained in the fall of 1960—but I had already taken the exam and was sure I had passed.

I was essentially the structural engineer for the Marquis and Stoller firm, operating out of a room in my house. My first project with them, the St. Francis Square housing complex, took me about a year. About three months before it was complete, a light bulb went on: I realized that I didn't have another job lined up. So, I visited some architect colleagues of Marquis and Stoller. Bob and Claude had given me very good press, so I got some jobs that way, one in San Luis Obispo, a courthouse. There was never a problem with those first projects, which helped my reputation. Those were the days when the engineer could walk on the site, tell the contractor what he had done incorrectly, and then the contractor would thank you for the correction and get it done right. No nonsense then.

Reitherman: In one photo of you when you were a structural engineer on your own, it shows what must be a T-square.

Elsesser: Yes, parallel rules had not yet come in. That was how you drafted in those days. I could turn out two or three sheets a day.

Kang: How do you see the engineer's role today?

Elsesser: It's totally different. The engineer is pushed around like everyone else. There's no leadership on the job. It tends to be run by the contractor, which is unfortunate, because it affects quality.

I did some very interesting structural work for Campbell and Wong. Worley Wong was the key designer there. I was still doing almost all the structural work for Marquis and Stoller.

Finally, after about a year, I rented a small amount of space next to the Marquis and Stoller office. It was near the Buena Vista Cafe, known for its Irish coffees, at Hyde and Beach Streets. I had one draftsman, who started part-time, and soon I hired him full-time. Then there was a part-time secretary, so the firm was now up to the size of about two-and-a-half people. Then, some neighboring space opened up and the firm doubled in space, to an area about 20 by 20 feet. By then, the firm was four and a half people, and at that scale, I ran the firm myself.

Marine Projects

Elsesser: In my early years of practice, in 1965, we got involved doing some engineering for Healy Tibbetts, a marine contractor, then the biggest on the West Coast, who one day asked whether I would be interested in coming to a meeting the next day to discuss their structural engineering needs. So, I went over to Stacey's bookstore and bought six books, went through them that night, and showed up the next day saying I would take on their needs and would answer their questions.

Kang: An overnight expert!

Elsesser: Yes. I think they had been talking to 500-person firms who had been charging a fortune, and they thought they could get straight to the expertise they needed more efficiently. This owner was the cheapest client in town. He would always buy used equipment, never buy it new. When he came to town, I arranged to take him out to dinner at the World Trade Club at the Ferry Building. As we were driving there, he called out, "Stop!"

He spotted a hamburger place, and wanted to quickly have some food there then get back to work.

For their waterfront and offshore projects, Healy Tibbetts would accomplish the kinds of excavation and foundation work a contractor would do on dry land, but using their pile driving barges and other waterborne equipment. My office at that time was two engineers—Fred Willsea and me—and two drafters.

One of our earliest challenges was a rolling mill order that had to be placed in Japan within an hour. We had to estimate what would be needed for all the piles at two sites—thousands of feet of piles. We estimated the piles and added something like 10 feet per pile. We had virtually no geotechnical information to go on—we just knew it was sand sitting in seawater. A year later after all the piles had been delivered and driven, we were within 20 feet on the total material required. The job worked out very well.

That project involved a salt transfer facility at Scammon's Lagoon in Baja California. Salt was barged sixty miles out to sea to load onto 1,000-foot-long bulk carriers. It was a very complex industrial facility. They had their own plane and flew us there to a landing field in Mexico. With the crosswind, we landed on one wheel until we rolled to a slower speed and got on two wheels.

Kang: That was a challenge. Would you take on a job like that again?

Elsesser: Sure. If you knew your structural engineering, you could go through a good textbook and understand it. The Internet today is more scattered.

In San Francisco, we did the engineering design work for Pier 8 and the new ferry terminal facilities at the Ferry Building.

In 1969, my structural engineering firm had four and a half people. That's when Nick Forell and I joined up.

Chapter 5

Forell/Elsesser Engineers

We started the firm with a year of engagement, followed by the actual marriage, as Nick put it.

A Trial Year of "Engagement"

Kang: How did the Forell/Elsesser firm start?

Elsesser: We started the firm in 1969 with a year of engagement, followed by the actual marriage, as Nick put it. [Laughter.] We had a year to see how we worked together, to decide if we were really going to form a solid partnership.

Kang: How did you two meet?

Elsesser: It was through common architects, but specifically it was probably through Chester Bowles on some school projects.

Kang: Where did the new Forell/Elsesser partnership locate?

Elsesser: Nick had office space at 50 Green Street in San Francisco. An architectural firm had moved out and he had extra space. I had a staff level of four and a half people, and he had about seven. So we moved in together.

Kang: This was after the trial year to see if the partnership made sense to both of you?

Elsesser: No, this was during the trial year. I ran my staff and office, and he ran his, but we were in the same space. You could say that during the "engagement" we lived together. [Laughter.]

Early Years of the Firm

Elsesser: We knew we wouldn't agree on everything. Disagreeing shouldn't be enough of a reason to end a partnership. We decided that if one of us wanted to do a project and the other one didn't, you could go ahead with the project and it was your problem if it didn't work out. We agreed to always compare notes on the important engineering decisions on all the projects.

Soon after we joined in 1970 to form one firm, the bottom fell out of the construction industry. We had to go hat in hand to clients asking for work. Those ups and downs have never been a surprise to me. It's the nature of the business. Later, construction picked up, and at that point, my earnings went up considerably from what they were when I was an employee at John Blume's firm. Then I gave my employees a 100 percent bonus—that is, their compensation doubled.

Kang: How did you and Nick set your fees?

Elsesser: Some architects would come to us and ask us for a price. Nick and I would look at each other and realize the client was primarily interested in price, not quality. We would quote a fee that was more than the architect was getting. We didn't want those jobs.

Kang: Who was the first engineer who worked for you?

Elsesser: Ephraim Hirsch.

Kang: No kidding! You go way back.

Elsesser: Yes, Eph Hirsch, a good friend of mine, was my first structural engineer. Of course, you know Eph for his subsequent career and his own practice in San Francisco. After he left, he was replaced in my firm by Fred Willsea, another good engineer. Fred and I had worked together for Blume, and Fred had earlier been a newspaper reporter.

Kang: Explain how you approached your structural engineering projects?

Elsesser: I never developed an analytical approach to engineering.

Kang: But in the Blume office you were given some of the most challenging analysis tasks.

Elsesser: I was always able to do the analysis in my career, but I had a conceptual approach rather than an analytical approach. I didn't start a structural design that way. In this early period of my career, I did over a hundred houses for architects, some of them rather challenging—hillsides, unusual layouts. You need a very clear idea of what you were trying to accomplish and quickly get to the structural solutions you would use, then refine them with analysis.

We did a lot of public schools. As you know, we have special code regulations, especially for seismic, in California. In addition, when the contractors' bids came in, if the low bid was a penny over the pre-determined budget, the project didn't go forward without re-design. You had to have the right design concepts. The schools had to be minimal and low cost, but not low quality. Sketching out design concepts at the outset was important.

We would sit down with the architects and talk the project through before we touched pencil to paper. The good ones loved that, bouncing ideas back and forth. The bad ones wanted to complete their architectural work and throw that design over to you to make it work structurally.

Working With BSD

Elsesser: In the early 1960s I started to work with BSD, Building Systems Development, a good-sized architectural firm in San Francisco. We can touch on a few specific projects later. Ezra Ehrenkrantz headed up BSD and also was a professor in the architecture department at Berkeley, and the number two man running the firm was Chris Arnold. They were doing marvelous studies of building systems. I turned out hundreds of structural variations—if you don't consider them all, how do you know if you are selecting the right one?

They initially developed a prefabricated system of major components that was used for a number of high schools in California. Ezra was a great marketer and there was a lot of work. Later, he moved east to set up an office in New York, while Chris stayed in San Francisco and ran that office.

The Building Systems idea was not just how to efficiently coordinate competing criteria to create an efficient design—it was to really understand how the building would function. I had a lot of projects with BSD. They were great fun.

Kang: What was driving this line of work? Was it an economic trend?

Elsesser: It was Ezra Ehrenkrantz. People would listen to him, and he could convince them of his ideas.

Kang: Who funded all this innovative research and design?

Elsesser: Primarily the federal government and universities, in particular the University of California and Indiana University. In those days, there was more money for such things, before they bought a war, the Vietnam War. When else did we have an opportunity to investigate 6,000 framing systems? It was a great time.

Kang: Has the systems approach worked its way back into architecture and engineering?

Elsesser: Most people don't take the time to work out all the combinations. They consider half a dozen alternatives and think it's enough. It takes a lot of effort to investigate all the alternatives, the thousands of ways to frame a building. One system can turn into another when it's inverted, and so on.

That project changed my career because it forced me to consider all the structural options and combinations. You could come up with 6,000 combinations of ways to span space in a building. You wouldn't use 90 percent of them, but they were all things that had been used or were possible.

Working With McCue, Boone, and Tomsick

Reitherman: You worked with the McCue, Boone, Tomsick (MBT) architectural firm a lot, didn't you? What was it like working with MBT?

Elsesser: It was an interesting office. When I moved my office from 737 Beach Street in with Nick's, Nick was in a building on Green Street—the Forell/Elsesser firm was launched there and stayed for several years. A couple of other good architectural firms had also been in that building with MBT. One was Rockwell and Banwell, partners of John Lyon Reid and Alex Tarics. They lost their shirts on a job and had to move out. That was when Nick acquired more space in the building, which was the space that I moved my firm into during our year of "engagement." Then we moved to Sansome Street and had a second-story space there. After that, around 1978, we moved to the building on Clay Street where MBT was.

MBT was really run by Rosalyn Koo, "Roz," an Asian woman who marketed the firm to clients before the client even knew they were going to need a new building.

Every Friday they had an office-wide crit session to go over the designs. Gerry McCue, even when he left to head up the Graduate School of Design at Harvard, flew back to San Francisco once a month for these sessions.

Continuity of the Firm

Kang: If an engineer were to come to you and talk about starting their own firm, what advice would you give?

Elsesser: If you're really creative and want to do interesting things, go for it. It's not a passive business. The people who do well, financially and otherwise, are aggressive. They go out and try to get the kinds of projects they really want to work on. If you want to be a partner in a firm, you have to work toward that goal and want to take on responsibility.

As the firm grew in the 1980s, we picked up as principals David Friedman, Jim Guthrie, and Bill Honeck. There were other people who didn't want the responsibility. Not everyone wants to be responsible for an engineering firm. David has been president for a number of years. We always had a president, but in the early years it alternated, every year, between Nick and me. Then one year I said, "Nick, you like being president and doing that job. Just keep that job and keep on doing it."

Kang: How did you retain your top people? When I was made a principal, in 1995, the principals were: Elsesser, Forell, Honeck, Guthrie, Friedman, Elizabeth Halton, Simin Naaseh, Paul Rodler, Mason Walters, and myself. Mark Jokerst was also a principal then, and has since retired. We were a big group of principals.

Elsesser: We welcomed that development.

Reitherman: From your point of view, Grace, it is a big achievement to be a principal, but does some stress go along with that?

Kang: It's not an award, it's an investment you have to make in the firm financially and personally, and you take on an increased level of responsibility. You are making a commitment. It's a marriage in a business sense. You have to communicate not with one person in this marriage but with several people. One of our successes has been to be able to communicate well with each other. We all love what we do. We have a passion for what we do. We have a sense of ego for the firm, not the individual. The overriding philosophy is to consider what's best for the firm. And we want it to go on. We've seen what Nick and Eric have done with the firm, and how in recent years David Friedman has guided it as president.

We all have the same challenge. What could we be doing better? We think beyond the structure itself in that regard. We feel that this approach is also the way of the future.

Arnold: There you go, Eric. You've succeeded. Eric, I think your firm has the future you intended for it. If your firm has a defined philosophy, you tend to obtain the right kinds of people—they find their way to you.

Elsesser: The big old firms in San Francisco have all vanished, like Brunnier. The H.J. Brunnier firm used to do most of the big projects here from right after the 1906 earthquake on up to the 1960s, around the time when the Bank of America building went up. They didn't structure themselves to have continuity. The same was true of John Blume's firm. The Degenkolb firm is still in existence, which is an exception.

Reitherman: The Blume firm was remarkably centrifugal, spinning out engineers who went on to form their own firms.

Elsesser: Blume had some of the best engineers. Some of them stayed several years, but then went on to work elsewhere or start their own firms.[6]

6 Some prominent "alumni" of the Blume firm are listed in the EERI oral history of Joseph Nicoletti. *Connections: The EERI Oral History Series—Joseph P. Nicoletti*, Stanley Scott Interviewer. Oakland, California, Earthquake Engineering Research Institute, 2006, p. 106.

Chapter 6

Engineering Projects

We sought the best structural solution for each project. As a result, the history of the firm is parallel with the evolution of better and better structural systems.

Northridge Medical Arts

Elsesser: One of the most challenging architects I worked for was Paffard Keatinge-Clay. He came from Europe, where he had worked for some prominent offices including Le Corbusier. Here in the U.S. he had worked for SOM in Chicago. His office was across Beach Street from my first location near the Marquis and Stoller office, and I bumped into him at lunch.

In the early 1960s I worked on some small, modular medical buildings with him, very minimal and simple, but very elegant. One of them was the Northridge Medical Office building.

I would walk into his office and he would have a model of a building that might have four columns at the mid-point of four cantilever beams in a pinwheel configuration. You can make that stand up, but only when all four columns and their beams are all in place—so the construction process has to be carefully worked out. I discussed with him moving things slightly and it became much more feasible.

In the Los Angeles area in Northridge, we did a fairly large three-story medical building for Keatinge-Clay, Northridge Medical Arts, built in 1966, before I partnered with Nick Forell. Keatinge-Clay collaborated with Dion Neutra, son of Richard Neutra, on the design. It's a long building with reinforced concrete shear walls at both ends. There was a big cantilever over the entry, about 25 feet square, with only one column at the edge of the building supporting it. The column base was part of the foundation mat that extended out in front of building, under the canopy.

Kang: So the foundation system echoed what was going on above ground?

Elsesser: Yes. And it has gone through two major earthquakes without significant damage—the 1971 San Fernando and the 1994 Northridge earthquakes—whereas the two buildings across the street collapsed, twice.

For the design of the shear walls, I took the reinforcing bars from 30-foot-deep caissons and ran them, without splicing, two stories high into the wall. The idea was to make the wall an extension of the ground, so there was no amplified acceleration, and it would move like a rigid body.

French Hospital

Elsesser: Another project in the late 1960s with Keatinge-Clay was the French Hospital in San Francisco on Geary and Sixth—two large buildings with a large garage. It had a two-story underground parking structure with large bays, post-tensioned, which is difficult. There were big cantilevers and a ramp that penetrated the structure. We couldn't have columns outside, so we built this big pier, and then four stories of steel on the top. The transfer system had 80 tendons in every single bay, both ways—160 tendons crossing each other to make the transfer.

What was really interesting was that the medical office building on the site was composed of nine bays, 60 feet by 30 feet, 180 feet by 90 feet overall. We used four big piers, cores that went up through the building interior. The columns on the exterior were not designed to take the lateral loads. And I said, "The last thing I want is for one of those columns to break." So I put in rotation and sliding bearings to make sure the system behaved the way it was intended to. It was really so easy, when you think about it.

Kang: How did you spec out those bearings?

Elsesser: We talked to the people around the country who made them. They're not common products.

San Francisco State University Student Union

Elsesser: Ultimately, we did the Student Union at San Francisco State College, now San Francisco State University. Design started in 1969 and the construction was finished in 1975. That project aged me ten years. There was a big legal dispute. There were initially about fifty architectural drawings, and then Keatinge-Clay proceeded to generate over a hundred additional sheets, and he expected us to take the European approach, where the contractor would just adapt to that as construction proceeded. It was a big mess.

Kang: Some of the Forell/Elsesser projects, like the Student Union, have been retrofitted or remodeled by other firms years later.

Elsesser: That's the way it often goes. And often the other firm doesn't understand the original design concept and makes things worse. The Student Union at San Francisco State, designed for seismic forces two and a half times code, was fine, but they hired a firm to retrofit it that did not understand the structural system, and they had to shore it in places they didn't expect.

Other Early Forell/Elsesser Projects

Elsesser: While the San Francisco State University Student Union building was a difficult project and a headache for us, it led to SOM knowing what a good job we had done and asking us to do the initial structural design of Davies Symphony Hall. They had a fabulous staff. The curving staircase on the exterior was a complicated architectural feature to work out. We were at a meeting when that design feature was brought up and we asked when we would get the drawings to figure out the engineering, and they said "Tomorrow." And they did it. Sylvia and I subscribed to second row center seats and have attended the past twenty-five years.

Reitherman: Wasn't one of the larger Forell/Elsesser projects in the firm's early years a high-rise building in Santa Rosa, California?

Elsesser: Yes, it was the ten-story Bethlehem Towers residential building, the tallest building between San Francisco and Portland. It was designed to have 8-foot, 6-inch floor-to-floor heights: 8 feet of space and 6 inches of slab. The beams were upturned along partition lines. We made it as simple as possible.

Reitherman: How come the upturned beam isn't used more to reduce story heights?

Elsesser: You usually get into conflicts with running the ductwork. But it's a good question. Maybe you can design the ductwork differently, or not have the ducts. That's the kind of thing that jumps into my mind. You pose a problem, and I try to think up alternatives. For me, that's what engineering is all about, doing it one better.

Seismic Reviews of Large Complexes of Buildings

Elsesser: We did an initial seismic evaluation project for IBM at their Cottle Road, San Jose campus in the San Francisco Bay Area. We had to quickly sort out the earthquake risks of about thirty buildings. We were able to point to the ones that needed further analysis and probably retrofit, and then a program was approved to proceed along that path. They pulled the drawings for all the buildings—hundreds of pounds of rolls of drawings, naturally. And in a day, Nick and I looked through them, toured the buildings, and gave them a quick report. It takes a lot of experience to look at construction drawings and extract the key information. We said, "How's this for a first pass?" and they said, "You're hired."

If you have enough experience designing buildings, and you flip through a set of construction drawings, the problems jump right out at you.

Kang: It must also require a visualization ability, to look at two-dimensional representations of the buildings, the construction drawings, and be able to see the three-dimensional structure and its flow of forces.

Elsesser: Sketching is related to that knack. Nick also had that ability.

After the basic structural upgrades were accomplished, we were asked by IBM to strengthen all the service systems in critical buildings, which meant strengthening and bracing all the utility lines.

We had a similar large project for TRW in southern California. In that case, there were eighty buildings on three sites. But as soon as they lost their big space contracts, they didn't care about the long-term protection of their facilities.

TRW had hired another firm, and for each of eighty buildings they had prepared a report several inches thick. TRW called us in and asked us to look through one. We did, and said, "It's a thick report that doesn't say anything." And they said, "Yes, that's what we thought." So we proposed to give them three pages per building. We started work for them, and it was what they needed. These concise summaries stated what the problems were, and what the seismic solutions were.

We went through the satellite assembly building, a very big building. It had three very important satellites being assembled. I looked up and said, "That light fixture is right over this satellite and might come down in an earthquake." The next trip we made, they had taken down the light fixture.

Another project where we reviewed a large collection of buildings was for the General Services Administration, evaluating the seismic capacity of more than thirty important federal buildings. The reviews led directly to final structural solutions, some done by our office, some by other firms.

We also reviewed over twenty buildings for Pacific Gas and Electric. Besides that broad review, we were the structural engineers for the retrofit of the PG&E Building in San Francisco after the Loma Prieta earthquake. That is a tall building dating back to 1924, and we had structural and nonstructural challenges. The treatment of the terra cotta exterior of the building was a big challenge, to preserve it. But it had been damaged in the Loma Prieta earthquake. I realized that in a contemporary building you have cladding that is jointed frequently. So we had them cut 3/8-inch joints through the cladding at every story level and seal the joints with non-rigid material. The cuts follow existing mortar joints, so you can't see it.

A project we did for Chevron was a large building complex. MBT was the architect, Nick Forell the Forell/Elsesser lead engineer. One of the buildings, the computer center, was especially vital, and we used eccentric braced frames.

San Francisco Museum of Modern Art

Kang: Eric, can you mention an architect you have enjoyed working with recently, when you had the opportunity to integrate the architecture and structural engineering?

Elsesser: I enjoyed working with Mario Botta on the design of the Museum of Modern Art in San Francisco.

Kang: How did that job come to Forell/Elsesser?

Elsesser: We were called one day by HOK [Hellmuth, Obata, and Kassabaum], the architecture firm in San Francisco. I didn't know anything about the project except that the Swiss architect Botta was involved. I had a book on Botta. In fact, Nick and I always subscribed to the architecture journals to keep abreast of that discipline. So we went over to the HOK office to talk with Botta. We didn't do any kind of formal presentation; I just took the book and said I knew something about his architecture.

The building started out as a concrete building, but it started to become heavy-handed. We did the whole scheme that way, but said we wanted to investigate using a steel frame with brick panels. The building is simple on the outside but complicated inside, with no two floors the same. Botta said he had never done a steel building. He considered it and decided to go that way.

I think Botta understood more English than he let on, but he would speak in Italian and this was translated at our meetings when he would come over from Italy once a month. At one meeting, he went on and on, very animated, raising his voice, gesturing wildly. When he stopped, we asked the translator what he said, and the translator replied, "He is very angry."

Botta had always used brick, and he was determined to always use brick in the traditional way, one brick on top of the other, forming bearing walls. We suggested that instead brick panels be hung from the steel frame, and eventually he came around to that solution. Botta designed every detail in the building. It's a very finished building.

Arnold: That's a European approach, to treat the building as a work of art. Like Arne Jacobsen, the Danish architect, who when he designed a college in Cambridge, England, designed the flatware as well as the building itself.

Elsesser: I visited Botta's office in Lugano. You walk four stories up a stairwell with a dim light bulb at each landing. You open the door and it's full of young people, fresh out of school. He apologized for not having more time, but they had to finish and deliver to France the next day a model of their design of a cathedral—and they had barely started it. [Laughter.] That's the typical charrette approach.

Ford Assembly Plant, Richmond, California

Elsesser: The Ford Assembly Plant in Richmond, California, was an extremely light building. We reinforced it with cable braces in both directions.

Kang: Wasn't there another scheme that required extensive foundation work?

Elsesser: Yes, but they couldn't afford that solution. In this building, the columns are very light and were designed as cantilevers. They support very little load. The columns had to do all the seismic work, but there wasn't enough capacity. So we put in a cable system and ran it both ways, and dispersed the forces so that there was minimal to no foundation work. So we said, "Let the building move," because the whole thing could move and the worst that would happen was they'd break some glass. And another engineer got a hold of the job after we had done this. And said, "Well, this requires a retrofit." I said, "No. It doesn't." I said, "If you wanted us to completely retrofit this thing, we would have had to replace these columns and put in a whole foundation system—a whole new structural system, throughout the job. And it would have cost about ten times more than what the budget was.

Kang: It was constructed with the cable bracing?

Elsesser: Yes, it was. My whole attitude was to improve the condition.

A Chronology of Structural System Development

Kang: Let's follow a chronological thread through different kinds of structural systems Forell/Elsesser has used over the years and have you explain what you have learned.

Elsesser: We sought the best structural solution for each project. As a result, the history of the firm is parallel with the evolution of better and better structural systems.

We know a building with just shear walls will end up with damaged shear walls in the big earthquake, unless it is low-rise, lightweight, and has lots of walls. We know braces eventually are damaged as they respond to increasing forces. In contrast with those older systems, we know that seismic isolation works and works very well. We know dual systems work. Moment frames work well if there are enough of them throughout the building. Shear walls with connecting links, localizing the damage to the links, are a good solution. The engineer must determine where the inelastic behavior will occur, and, which is sometimes overlooked, consider how that deformation will affect the rest of the building. The reduced beam section solution to moment frame joint damage can propagate a deformation along the beam and cause other damage, for example. Locking up the nonstructural elements into the structure is asking for damage to the nonstructure.

The eccentric braced frame is a way of precisely locating the structural fuse, controlling where the damage and need for repair will occur. We used the eccentric braced frame for the San Jose Federal Building, the Federal Express headquarters building in Memphis, the UCLA Medical Office Building, the California State University Sacramento Library II Addition, the Chevron Park office buildings in San Ramon. We even hired Egor Popov on a couple of the jobs to make sure we were carrying out the eccentric braced frame concept. Another good system is the buckling-restrained braced frame (BRBF), which has an unbonded brace that is confined to prevent it from buckling.

Kang: The BRBF perhaps is better at localizing inelastic behavior so that it can be repaired as needed after an earthquake. With the eccentric braced frame, damage to the floor beam is a problem.

Elsesser: But with the BRBF, you need to take finishes off just to access the brace to inspect it. You can always argue you should have a dual system, a moment frame plus one of the bracing systems. That makes a lot of sense. Now you can build a tall building with just a concrete core as the lateral force-resisting system, but that's not a good system. A concrete building is always going to crack. The cracks can be small or large, but they occur in the earthquake and you have to access them to repair the damage. You want to stay away from relying only on shear walls unless the building is relatively small. And you don't want to rely on just a steel moment frame, except for a tall building. The ground motion has most of its intensity in the region of a period of half a second, while the very tall building may have a period of several seconds, and so the building won't respond much.

Kang: What about ductile concrete moment frames the firm has designed?

Elsesser: We used a lot of them, for the Cal Poly San Luis Obispo School of Architecture, the San Francisco Art Institute, and the French Hospital in San Francisco. They were well detailed and good buildings. But I think even those well-designed frames will still crack and be hard to repair, so we kept looking for what was better.

Kang: Then there is the steel ductile frame.

Elsesser: We believed in steel for a long time. The Northridge earthquake showed this system had vulnerabilities.

Kang: With current detailing, can't you design a good welded steel moment-resisting frame?

Elsesser: You can avoid the fracture problem, but you move the problem away from the joint into the beam and you still have damage that is difficult to repair—you have to take all the cladding off, for example. So that's not acceptable.

Kang: Don't we have the same issue with the distortion of the beam in the eccentric braced frame?

Elsesser: I think it performs better. But you still need to take the building apart to inspect the link beams to verify the performance, unless the building is somehow wired.

Kang: Then there's the category of cantilever columns, coming out of a grade beam.

Elsesser: In effect, the inverted moment frame. We did a lot of them for one-story schools and it is very inexpensive.

Kang: Now we come to the system called cluster moment frames.

Elsesser: A tower of moment frames, with collectors to bring the loads to them. Instead of distributing them throughout the building, you have clusters. If you can expose them, you can inspect them after an earthquake. If you have to fireproof them, you have a problem. We used this on the San Diego Intel assembly building, the Santa Cruz government center, and the Petaluma general mail facility. You have to think of the cost of the repair after the earthquake, not just how the building will perform during the earthquake. For the Intel job, we used a module of 40 feet, because decking came in that length, and the same sizes for all the columns, and also for the beams, to make it efficient. It was cheaper than a tilt-up.

Kang: Here's another category of structural systems Forell/Elsesser has used: steel truss moment frames, as were used in the Surge Lab at U.C. San Francisco and the Hewlett-Packard Assembly and Laboratory facility.

Elsesser: We used steel Vierendeel trusses instead of beams. It allows for renovation of the lab services for different projects in the case of UCSF. The Hewlett-Packard building in Santa Rosa also needed the same flexibility, so you could route the mechanical and other systems through them.

Kang: Describe the coupled wall systems the firm has designed.

Elsesser: We used that on the Life Sciences Building at U.C. Berkeley and the Physical Sciences Building at U.C. Irvine. The link beam is where the damage will occur, and that's acceptable. We also used it on the PG&E headquarters building, where we used other systems as well. They cascaded in their resistance to the earthquake. The coupled walls were stiffest and their link beams would go first. Then the moment frames would take up the slack. Finally, the other shear walls, designed to yield in their boundary elements, would then contribute more to resisting the earthquake. A progressive system.

Kang: In that project, we also joined some of the formerly independent buildings together.

Elsesser: Yes, we joined two L-shaped buildings and effectively made one U-shaped building.

Kang: And the entire system was joined at the base?

Elsesser: Yes. The building was mounted on timber piles, untreated wood piles, but upon investigation they turned out to be okay. We needed a few more piles, but were able to tie the existing piles to the new ones with a new pile cap.

In the design of the PG&E Headquarters Building in San Francisco, we used shear walls with link beams, moment frames, and big shear walls at the bottom of the structure. We got a huge amount of capacity out of the whole system. This was an idea that our friend Tom Paulay in New Zealand had. I had read of his work and waited until I had the opportunity to apply the idea. It was a three-part building system, like the one designed by Kiyoshi Muto in Japan, the Kasumigaseki Building in Tokyo constructed in 1968, the tallest building in the country at the time.

Limiting Damage

Elsesser: In our design for the PG&E Headquarters Building, we had concrete walls that were designed to be fractured, but easily repaired. Then we combined two other systems, to get them to work together. People like Muto and Paulay were really trying to think seriously about the basic problem.

PG&E also wanted to have drawings showing what the anticipated damage would be after the earthquake, damage that would not make the building unsafe and require it to be closed. So we did drawings showing where localized cracking and damage would occur.

Kang: The trend these days is to use extensive analysis to justify initial premises. Two engineers may have two very different opinions as to how a building will perform in earthquakes, and both can get their analysis to demonstrate they are right.

Elsesser: Which may mean they are both wrong.

Kang: Would you design some of your early work differently if you had it to do over today?

Elsesser: We, like lots of engineers, designed some reinforced grouted brick bearing wall buildings in the 1960s and 1970s. You can design them well, but eventually you come to realize that in an earthquake the walls will crack, and when they crack, you have expensive damage to repair. That's an example of a system I wouldn't use today. If you have enough walls in a low-rise building—the classic case is a prison—you can use a bearing wall system, but often with that system you will get damage that is costly to repair. Engineering evolves. You have to keep looking critically at your own work and keep improving it. It's difficult to keep changing, but that's the fun of it.

Structural Systems—Future Improvements

Kang: Don't we have to allow for future improvements? Designing for energy dissipation is a relatively new trend, but aren't there other new developments yet to come along? The industry that structural engineers are in is really thousands of years old. I am convinced there are further developments past our own time. What about computer form-finding, using the computer not only to do the calculations on a design but to work on optimizing the layout and create a building form, from architectural design to structural engineering?

Elsesser: You still need the engineer to really think about the problem. When you talk to the better architects and present such ideas, they may or may not buy your approach for a particular project, but they say, "Wow, you're really thinking about this design problem." They are impressed that you are thinking about the whole design problem. That's all the architect ever wanted to do was think about their particular project. If you think creatively, you'll get all the work you can possibly handle.

Kang: What do you think makes a good structural engineer?

Elsesser: It requires an exploring mind. You have to want to find out. If you simply accept the computer output, you're a technician, not a structural engineer. You have to test the design.

If we look at past earthquakes, you can see patterns of what works. EERI collects such damage information now, but nobody uses the information. Now that we have the computer, engineers don't think they need to look over their shoulders at earthquake experience.

Chapter 7

Isolation and Other Seismic Protection Innovations

As far as successful design for earthquakes is concerned, I think we've hit the top with base isolation.

Elsesser: As far as successful design for earthquakes is concerned, I think we've hit the top with base isolation. We've taken everything and put it in one device, and it does most of the earthquake work. It's the greatest thing the New Zealanders have ever developed.

Kang: Is this solution ever *in*appropriate?

Elsesser: When you have buildings right next to each other, as is common in a big city, you may not have room to allow for the movement.

Kang: What about a site prone to long-period response?

Elsesser: You wouldn't want to use isolation on soft soil in Mexico City. But here in San Francisco, even on Bay Mud, you can use it if you make sure you have a period shift between the ground's response and that of the building.

Visiting New Zealand and Learning About Isolation

Elsesser: I had the chance to visit New Zealand with Chris [Arnold] in April of 1984, on a project about protecting nonstructural elements from earthquake damage.[7] A couple of the earthquake engineers there later visited us in California. It was very efficient learning. We talked to the construction people in New Zealand, and they were really smart, as were their engineers. We saw what they were doing with base isolation there, the William Clayton Building in Wellington, the first isolated building in the world. That had a big influence on me.

Arnold: There was also the Union House building in Auckland, near the waterfront.

Elsesser: Yes. It was isolated in a different way. It has steel diagonals that come to a point slightly above grade, and that base level was attached to the ground laterally with steel bars. Below that level were long piles that were free to move in caisson sleeves. The steel bars dissipated the energy that would otherwise have gotten up into the building.

Here in this country, the first isolated building was the Law and Justice Building in San Bernardino, California. Alex Tarics was the engineer. Alex also tried to do a base isolated project here in San Francisco, but it never went forward. The U.S. is way behind Japan in adopting this innovation. There are about 3,000 isolated buildings in Japan now.

The design and construction industry wanted to go ahead and do it in Japan, and the owner could then charge higher rent.

Arnold: The problem is that in the United States, the public does not experience damaging earthquakes often enough. If the U.S. had three Loma Prieta earthquakes every decade, things would be different.

Reitherman: It's a testament to top seismic engineers like Eric in this country that they can stay at the top of the game with Japanese colleagues, when Japan as a whole country is exposed to so much more seismic risk. Earthquakes are a national problem and an influence on the national building industry in Japan, but not in the U.S.

Salt Lake City and County Building

Elsesser: The Salt Lake City and County Building was a retrofit project we got in the early 1980s, shortly after Nick and I returned from a World Conference on Earthquake Engineering. It had been known as a seismically hazardous large old masonry building for some time. Various consultants had looked at it, but no one had come up with an effective solution at reasonable cost. I was visiting Ezra Ehrenkrantz here in San Francisco, who was the architect for the renovation of the building. Salt Lake City had gone through four or five architects trying to find seismic and renovation solutions. I looked at the drawings and saw the big crawl space under the building and said it was ideal for base isolation.

7 BSD Inc., Forell/Elsesser Engineers, Inc., and KRTA Limited, *Seismic Design of Architectural Elements*. March 1987.

I went to Salt Lake City for a meeting, and they asked, "If we do this, will it put Salt Lake City on the map?" I said, "Of course," and they said, "Do it." We worked with Ron Mayes and his group at DIS, Dynamic Isolation Systems, and developed a seismic retrofit design that included base isolation. We solved all the new technical problems involved.

We had to tune the isolation system. We needed a certain amount of damping, and at first we designed all of the isolators to have damping, but that didn't work out well. Then Ron suggested putting high-damping isolators only around the exterior and isolators without the damping on the interior, a little over 200 of each. We went round and round to balance the system, repeated analyses. We needed to reinforce the base around each isolator with some significant foundation work.

We then needed to figure out how to cut the building loose from its existing connection with the foundation and put the load on the isolators. The contractor said, "That's easy. We'll use a band-wire saw." We said "What's a band-wire saw?" He explained it was used in mining—it's like a band saw. We also used some conventional steel retrofitting up in the tower, but essentially isolation was the answer.

After that, we collaborated with the office of Larry Reaveley and his brother to do some other projects in Utah. Reaveley has an interesting custom in the office to show movies once a week, but they turn off the movies promptly at 1 pm to get back to work. The movie would of course last longer than that, and they would show the rest of it the following week.

We also did work on the Utah State Capitol, as well as the Salt Lake City and County Building. I think the Salt Lake City project really changed the character of Forell/Elsesser. It made us want to climb the ladder and keep on climbing, finding new things.

The Salt Lake City job put us on the path to always seeking out new improvements. When the project was being done, the firm spent about $75,000 in billable hours as Mason Walters developed the system from the "ground up," because the knowledge base was not prevalent in the profession. We received a fee, of course, but we invested our own funds to advance our knowledge. Ron Mayes and I did a lot of lectures together to familiarize people with the concept.

I used to say to people, "Can you imagine not having rubber motor mounts in your automobile and what the vibration would be? Can you imagine having the axles and chassis rigidly connected to the wheels without any shock absorbers and springs?" But we have rigidly mounted buildings in the ground and let them suffer the vibrations from the earth when the earthquake occurs.

To get change, you have to say things with great conviction. You can't just say, this might work, this might be a good idea. You have to be convinced and convey that confidence to others.

Oakland City Hall

Elsesser: We were reviewing the existing San Francisco City Hall Building structure for the Building Department around 1988. Although we started to perform a review function for

San Francisco City Hall before doing anything for Oakland City Hall, we actually started the retrofit design work for Oakland first. That's because when the Loma Prieta earthquake occurred in October 1989, San Francisco City Hall remained occupied, whereas Oakland City Hall was evacuated immediately and remained vacated. Oakland was impressed that we already had a base of expertise from our work for San Francisco. The two projects somewhat overlapped, and both were monumental city halls with base isolation solutions. Oakland was able to obtain FEMA funding for their damaged building before San Francisco could get funding. We started the Oakland City Hall retrofit design in 1990. At the same time, we did quick damage assessments for San Francisco, which eventually evolved into the seismic retrofit project, "Phase 1," which started in 1992. We were prime on that project, with all the consultants working under our umbrella. "Phase 2" was led by Heller Manus and involved the interiors and renovations. Construction started in 1996 and was complete in 1999.

We were interviewed for the job of retrofitting and isolating the Pasadena City Hall in 1992, prior to the start of construction at San Francisco and Oakland. Pasadena had earlier studies done on their building, and they wanted another evaluation. We were awarded the contract for that project, with the Architectural Resources Group as the architect, in 1994, right after the Northridge earthquake hit in January. Pasadena City Hall was only completed this year [2007].

San Francisco City Hall

Kang: Tell us more about the San Francisco City Hall seismic retrofit?

Elsesser: When the building was designed in the teens to replace the one destroyed in the 1906 earthquake, it had all the wrong seismic features. It had a tall and soft ground story. That might prevent some damage to the rest of the building in a very small earthquake, but could make it collapse in a sizable one. The building was significantly damaged in the 1989 Loma Prieta earthquake, which was over fifty miles away. Meanwhile, across the Bay, the City of Oakland had closed their historic City Hall right after the earthquake, and that made getting federal money to really fix the building much quicker.

The only solution to historically preserve the big San Francisco building was to isolate it, to put 530 isolators under it. It took two years just to install the isolators. We had all sorts of interesting problems to solve, but nothing significant, no claims or disputes. The dome landed its weight on columns, but the walls were offset a full bay from the walls that carried vertical and lateral loads on down from there. We put a large space frame around the dome and tied it to the walls.

Kang: How did Forell/Elsesser get the job in the first place?

Elsesser: Ron Mayes was always curious about the building. He wanted to see the drawings and the city wouldn't let him. This was a couple years prior to the 1989 Loma Prieta earthquake. Other city buildings had been looked at, but not this one. Ron and I had been working to get seismic isolation adopted, but from the time I got back from New Zealand it took half a dozen years—until the Loma Prieta earthquake—to get things moving. Finally, he got someone to bring out the City Hall drawings. We said it should be isolated, but nothing was done. Then the earthquake occurred. They called us, because we knew something about the building.

The damage was more serious than was apparent. FEMA was proposing to provide a couple million in disaster relief funds to patch the damage. I said, "That won't solve the problem. You have collapse mechanisms here." There were multiple reviews by FEMA consultants who argued that either the damage could be patched, or the city should knock the old building down and start over. We told them the City Hall had more structural vulnerabilities than they had recognized, and that it had the same top category historic registration as the nation's capitol. I had been to Europe and photographed some of the great domed buildings there, like in Paris, and I said that San Francisco had one of these world-class buildings, and it had to be preserved. Each time, this review process took about a year.

The project proceeded through the terms of several mayors. Former San Francisco mayor Dianne Feinstein, a U.S. Senator by then, visited one day, and when she was briefed on the building and the project was just barely getting going, she said, "Get on with it—move people out as needed to get the construction going."

Our contract called for us to have two people on our staff there at the site all the time, something we insisted on. There were change orders and additional costs with regard to remodeling and building interior work, but not the structural work. And in this case, Forell/Elsesser was the prime, instead of the usual situation where the architect is the prime and retains the structural, mechanical, and other consultants. When our original scope with a budget of $104 million began to look like it would cost $140 million, I called all the consultants into our office and said, "We are here to remove $40 million from the project." In the effort to do the right thing and upgrade old services like HVAC and lighting, the cost had climbed. So we took all those upgrades out, and the city came to see the difference between an old safe building and an old, *modernized*, and safe building and agreed to spend about another $100 million on the modernization.

Kang: I recall that one of the Forell/Elsesser people there on site from our office was Paul Rodler.

Elsesser: Yes. Finally, the contractor and subcontractors realized Paul knew the building better than they did. They would line up to ask him questions at eight o'clock in the morning. I told him to answer the questions if possible without going back to the office. And he didn't need to write up everything, just get the job done. We had already done all the documentation necessary. We didn't like the approach used by some other offices of preparing working drawings that didn't think through and detail everything. The contractors realized that if they got the correct answer right away, they could go on with their work and it wouldn't cost more.

We also got the contractor and the City of San Francisco to split the cost three ways with Forell/Elsesser of having Bob Canfield there frequently to photograph the project. He was originally an architect we had worked with. For some of the shots up in the attic, it would take him eight hours to set up lighting. It's a great visual record.

Tony Irons was the San Francisco's key representative. He has a marvelous personality. The city didn't hire a separate architect, the city took that role on itself. We got on very well with Tony and worked out everything in advance. We had weekly meetings with him and our design team and spent time just talking about what might go wrong and prevent those situations. Steel members had to be brought into the building through a window, for example.

Kang: Did you consider alternatives to isolation?

Elsesser: We looked at the alternative of taking down most of the 300-foot-high central dome portion of the structure and putting a new one back up. Or put so many moment frames in that you tore apart the whole building to install them. We ended up tearing down brick walls around light courts and replacing them with reinforced concrete. Isolation saved the city lots of money. Opposite City Hall at Civic Center, across the plaza, is the Asian Art Museum, the former library, which we also isolated in a later project.

Explaining Seismic Isolation to Clients

Arnold: I have been very impressed with the level of professionalism of both architects and engineers in New Zealand. We had a nice tour by an architect who showed us an innovative building in which the walls were designed to rock in an earthquake. I asked him how he explained that to the client and laughed and said, "Oh we didn't worry the client." [Laughter.] How do you communicate to your client that their isolated building is going to move around relative to the sidewalk during the earthquake?

Elsesser: A typical case might be to design for roughly 30 inches of displacement each way, recurring over a time span of about 3 seconds. That's quite a ride. We don't actually try to explain the "ride" of their building. We may have to censor this part of the interview so as not to alarm clients. [Laughter.]

Arnold: At an EERI Annual Meeting, I recall saying that the ideal vision for earthquake engineering would be that in an earthquake, you and your building go for an exciting ride, but nothing gets hurt.

Elsesser: I think that ideal is best realized by isolation. You try to take all the motion out, as much as you can, at the base, where it originates, and then you don't have to deal with large interstory drifts up above. We really need to have a course in mechanical engineering taught to structural engineering students to educate them about isolation, damping, and mechanisms for dissipating energy. When I say "mechanism," structural engineers will think of the negative aspect—too many of those in a story and you have a collapse. But I mean the controlled motions at various kinds of joints and connections that mechanical engineers design.

Kang: A deliberate mechanism, with a positive purpose.

Elsesser: If every structural engineering

student were taught a course in seismic isolation, they would all do it. It is really a very elegant concept, and it works.

Reitherman: Explain to me why we don't isolate houses, at least some custom-designed houses?

Elsesser: If you designed the dwelling from the beginning, it wouldn't be a large cost, but the house-building industry doesn't want to do anything different. In China, they have built some houses with double slabs, with a layer of sand in between. There may be some simple solutions. The house may end up askew after the earthquake and you would have to re-locate it.

Arnold: Even if you wanted to make a little change in the way the house builders put the sill plate on the foundation, it would be a big issue. The developer, or the architect, usually hires the engineer just to get past the structural issues and go on with the project.

Elsesser: For a lot of houses, even on hillsides, you can extend the ground up into the building, dynamically speaking. You make the foundation, and a lot of walls, be an extension of the ground, without amplification. You can resist those forces with extra shear walls. If you just put in the minimum braces, you don't do that.

Kang: Why is it that isolation schemes are not more economical?

Elsesser: When the first isolation code provisions were written, you had to meet the conventional seismic regulations, as if it weren't isolated and it would receive the full force of the earthquake, as well as do what was necessary for the isolation systems. Some of the people involved in the review process were thinking of nuclear plants, putting a whole extra layer of requirements and conservatism on building isolation systems that didn't apply to the conventional design of the same building.

On the San Francisco City Hall project, we had one peer reviewer who did not want the project to go ahead unless it was "magnitude 8-proof."

Dual Systems

Elsesser: In cases where isolation is not the best alternative for some reason, I prefer dual systems. Engineers in recent years have begun to switch to the braced frame—I should say they are switching back to the braced frame. But we already learned that if you just have a single system, it is hard to make it work. We've learned that is a problem—that in a big enough earthquake, the braces will be broken. There's no place to dissipate the energy. A buckling-restrained braced frame, a BRBF, is different, because it will accommodate a change in length of the brace.

For some lightweight, low-rise buildings, you can get away with a single system, like braced frames, because their strength is large enough compared to the load. But when you get to about four stories, or heavier construction, that strategy doesn't work anymore. You're going to get damage. The earthquake is unforgiving when it makes the brace go inelastic. You get buckling, connections break, you get damage that is difficult to repair, you can even get collapse.

It's not just reliance on braces, it is a failure

to learn from past decades. When you rely only on concrete shear walls, you have to start by thinking, "If the building moves enough in a strong earthquake, it will crack. Where will it crack?" I asked that of the designers of the Rincon Towers buildings—the 49-story and 60-story projects underway by the Bay Bridge—where would the concrete crack? The answer was that the core would be damaged from the base up through the lower third of the building. That's a lot of damage, a lot of repair work, a lot of disruption to the lives of all those residents after any major earthquake.

Kang: The most common dual system has a reinforced concrete core with a perimeter moment-resisting frame, which qualifies for a lower base shear than if there were only the shear walls. The core is designed to take all of the lateral force, and then—in addition—the frame is designed to take 25 percent as a backup system.

Elsesser: We can call it a "backup" system, but the idea is to get both to participate, the frame starting to take up slack as the walls are cracking. The trick in a dual system is compatibility. You have to conceive and analyze how the different systems share the load as the earthquake continues. An eccentric braced frame and a moment-resisting frame can be intelligently combined, for example. The eccentric braced frame, of course, brings to mind the wonderful Egor Popov, who was able to do advanced research that could be applied by practicing engineers.

Kang: In the FEMA book [8] that Chris was the chief author of,there are charts in Chapter 7 that explain what combinations of systems can be designed to provide good seismic performance. There is also a graph showing the need for energy dissipation.

Elsesser: You need a system that can dissipate energy, without failing.

Arnold: Is this analogy valid? Somebody is putting bricks into a box you are holding. As the box fills up, you will eventually collapse, unless you can begin to hand off the incoming bricks to someone else.

Elsesser: The question is, where do those bricks go? How are they passed on to someone else? The energy doesn't vanish, it has to go somewhere.

Reitherman: The horseback rider can use the muscles from knee to hip to flex and raise the body a little to avoid the upward bumping from the saddle, and then gently lower the body to just barely touch the saddle, instead of falling on it. The work done by the muscles dissipates or handles the energy of those up and down oscillations that would otherwise be in the form of impacts. The horseback riding analogy comes to mind because of your diagrams of imaginative designs of entire buildings acting as if they were flexing their muscles, moving via mechanisms or joints, acting like giant springs.[9]

8 *Designing for Earthquakes: A Manual for Architects*, FEMA 454, Federal Emergency Management Agency, Washington, DC, 2006.

9 Eric Elsesser, "Seismically Resistant Design—Past, Present, Future," *Proceedings of the 13th World Conference on Earthquake Engineering*, Vancouver, Canada, 2004, paper no. 2034.

Elsesser: That's what the springs and shock absorbers in your car do.

Reitherman: Eric, during your consulting work with BSD in the late 1970s, which led to the building configuration book, you diagrammed how a building changes in an earthquake. It might start off with one pattern of response, but even if that snapshot of behavior were to be close to the response the seismic design was predicated on, as the earthquake continues you have a different building. Some walls soften before others, for example. You could get torsion or some other pattern of response quite different than at the starting point. Do most engineers visualize their designs in such a dynamic way?

Elsesser: No, most don't think that way. And in addition to the first earthquake, there is a high likelihood of one or more earthquakes, aftershocks, and one of those could be as big as—or even bigger than—the first one.

We looked at a telephone building in Oakland, Pac Bell at that time, which had been damaged in the Loma Prieta earthquake. We analyzed it to see how it would perform through another earthquake, and we showed how it kept degrading as it underwent more motion. Damage led to more damage. You would like the opposite, where some damage acts as a fuse and prevents worse damage. If you plot what happens as the building keeps cycling back and forth, and if over time the deformations keep on increasing, you're heading toward a very bad result.

You have to realize you are trying to predict upper and lower bounds of deformation and can't do it exactly. You may get one great jolt at the beginning of the ground motion that changes the whole pattern of ensuing deformations.

Reitherman: Tom Paulay says in his EERI oral history[10] that the one earthquake you know will not be experienced by the building you have designed is the one you have designed it for. If that precise design earthquake actually occurred, it would be a statistical fluke.

10 *Connections: The EERI Oral History Series—Robert Park and Thomas Paulay*, Oakland, California, EERI, 2006, p. 118.

Chapter 8

A Philosophy of Seismic Design

I never wanted to design a building that failed, and to me, it's a failure if your building goes through a major earthquake and can't be occupied within a month.

Kang: In climbing up to find new things, what new things do you see?

Elsesser: I think a whole new mechanism needs to be invented for how you want a building to behave in an earthquake. Just because engineers have a computer program to analyze a particular way a building behaves, that doesn't mean they have solved the problem.

Kang: What other ways might there be?

Elsesser: I see that the building has to move. Most buildings are not designed to move. I would say that if your floors are going to move, don't lock them all together and have all the structural and nonstructural problems of interstory drift. We could explore letting the floor move, relative to the whole building. I have presented some of these thoughts in my EERI Distinguished Lecturer talks.[11]

---

11 Eric Elsesser, "What's Around the Corner in Seismic Design," The 2007 EERI Distinguished Lecture.

The Nonstructural Problem

Kang: In the 1960s, the seismic provisions of the Uniform Building Code (UBC) were brief and not very detailed. Today, the UBC, the International Building Code (IBC), and the NEHRP Provisions all specify consideration for deformation and inelastic behavior and call for the nonstructural features, such as cladding, to accommodate the structural deformations.

Elsesser: The code basically treats the nonstructural elements as a contributor to the overall damageabilty of the building. I got Professor Vitelmo Bertero at Berkeley, an expert in cladding, and several other top experts to put together a proposal to NSF to investigate the use of cladding as a positive contributor to the solution of the damage problem, as a damper for a building. We got four great reviews and one negative one that sunk the idea. We were going to work out how to let the pieces rock and move in a beneficial way.

Reitherman: Didn't you work on one project with MBT that had a special seismic testing program to verify the design of the curtain wall? It sounded almost like special effects in the movies—first you rack the walls to simulate the earthquake, then you fire up a propeller-type aircraft engine and blow a gale at it, while also spraying it to simulate a driving rainstorm?

Elsesser: That was for the IBM facility in San Jose. Gerry McCue did one of the earliest NSF-funded earthquake studies of nonstructural components.

Then the IBM Santa Teresa project came along with steel frames that we designed for two to three times code force levels, but you still have interstory drift. When you get to the corner, you have one wall rotating one way and meeting a perpendicular wall rotating in a conflicting pattern. They designed a joint that would hold fast on one side of the building and break free in an earthquake on the other side. With a cherry-picker you would then drive around the building and snap the panels back into position, and it was weather-tight again. The buildings we did on that site have done well in two nearby earthquakes now.

Arnold: And the building had a large number of corners, with re-entrant corners in courtyards.

Elsesser: Gerry's approach was to develop several design alternatives and present them to the client. Nick Forell led this project for our firm. One alternative was a big simple box, one was very complex, and so on. The head of facilities said, "We'll take the big box." The head of the electronics line of work they were in preferred another design, the most expensive one, where every office was an outside office with its own windows. The offices were to be occupied by 1,500 PhDs they had hired and wanted to retain, and if it took nice offices, then that was a legitimate business expense. At that time, IBM had a large store of cash. They eventually selected the latest moment frame construction, and it was tested and instrumented by Caltech engineers.

Configuration Irregularities

Elsesser: I have pushed very hard to get the seismic code to account for the big factors that cause damage, such as soft stories, torsional imbalances, and discontinuities. In the 1970s, analysis was getting much more sophisticated with computers, but that can sometimes only mean you have a bad design that has been well analyzed. I argued that the code should steer the engineer away from serious configuration irregularities. Slowly, that has gotten into the Uniform Building Code and later the NEHRP Provisions, but more needs to be done. Engineers can be a hundred years behind the times. I pushed hard on the SEAOC committees to get these big changes implemented. There should have been some simple prohibitions as to what simply should not be built.

What is Acceptable Performance?

Kang: How do you see the current computer ability to make a virtual building and simulate it?

Elsesser: It's awful. They're using the computer output as the final answer, whether or not the program can model the real problem. It's a failing when the engineer doesn't really understand how the building will behave and only understands the numbers or depictions the computer produces. When you're not dealing with earthquakes, you can get away with all sorts of flawed design concepts that can be made to work adequately. But with earthquakes, you have to push your structural engineering to the limit. I'm passionate about seismic design. I never wanted to design a building that failed, and to me, it's a failure if your building goes through a major earthquake and can't be occupied within a month.

Kang: You're talking about a performance level beyond a minimum life safety goal.

Elsesser: Absolutely. I don't think that level of performance is adequate. That definition of acceptable goes back to the first Blue Book editions. To me, the purpose of the building is to be used, and if it can't be used after the earthquake, it's a failure. I simply don't want to be associated with any failure in my life, it's as simple as that. To avoid that kind of failure you can't just run the numbers and prove a design meets the code. You have to select the structural system and its layout so as to get the most performance you can. Engineers try to minimize their designs. Instead, they need to optimize their designs. I know the code says that in the worst earthquake you can tear down most of a city afterward, but that doesn't make sense.

Kang: But how do you justify the extra cost?

Elsesser: In a seismic area like California, you need to start with a different cost basis. A seismically designed building will cost more.

Arnold: It looks like, historically, that you set out on an evolutionary trail to find better and better structural systems. That's not the usual pattern. Many engineers become familiar with how to design with a few systems and stick to that. You have mentioned your interest in keeping up with research, and it almost seems as if you acted as the implementation arm of the U.C. Berkeley engineering researchers. That's unusual. It must have reinforced what people like Popov and Bertero were doing, and it helped what you were doing. And you sometimes have an anti-academic bent! [Laughter.]

Building Models of Designs

Kang: How did you use models to develop a structural engineering design?

Elsesser: I usually made a physical model of our designs. There's one hanging on the wall in the office, which I made by soldering little metal pieces together, one floor level at a time, then propping them up on their vertical supports and proceeding. The next day I would meet with the architect and he would be surprised to see his building. I enjoyed it, and it helped the discussion with the architect.

Kang: You also did an unusual model of the Jewish Museum in San Francisco. You said you were going build a model, and in a week suddenly it appeared in the office.

Elsesser: This was the model of the twisting frame design. It was a box with slots to insert frame elements at different levels, to show how it went together. I have a little wood shop downstairs at my house. I have to say we got fired from that job by Libeskind, when the cost was too high. It was $34 million instead of $22 million. And he hired Ove Arup, and the revised design came in at $70 million, and nothing happened for a decade.

Arnold: Architects always build models, but engineers don't seem to.

Elsesser: That's right, they don't. I never could understand how an engineer can solve a three-dimension problem without building a model. Beside the models for the design projects, I used to make a little model in my wood shop at home every year for each of our architect clients as a holiday gift.

Kang: In the Forell/Elsesser office, Mason Walters recently constructed a wooden stick model of a pedestrian bridge and brought it in. He could then show it to other engineers here to discuss the design and decide how to do the analysis.

Reitherman: Eric, could you mention the story of a full-scale model you made of a steel moment-frame joint? I think it was the Bethlehem Steel Building, at the corner of California and Davis Streets in San Francisco, where the beams were attached to the sides of the protruding exterior columns, setting up torsion in the joint. You once told me the story about how a model you made affected how the structural analysis was done.

Elsesser: That was when I was working at Blume's. I argued we had a problem with the joint, and no one believed me. I bought some Plexiglas and made the model at home, brought it in, and we looked at each part of the load path and what it did to the joint. The model convinced the office we needed to design special connections.

Arnold: That joint had its inception in the architecture, to express the vertical lines of the columns.

Elsesser: The San Francisco building was copying the design of the Inland Steel Building in Chicago. Welton Beckett, architect of the San Francisco Bethlehem Steel Building, wanted to have his go up first, so the design process was rushed. But the Chicago building had no significant seismic hazard to contend with, whereas we had to worry about that.

Arnold: The conceptual designer at SOM for the Inland Steel Building was Walter

Netsch. The building has been beautifully rehabilitated and well taken care of. Subsequently, Netsch did the conceptual design of the Zellerbach Building in San Francisco. He was one of our advisors on the early BSD project to design a system of prefabricated components for California schools.[12]

Building It Quickly

Kang: Eric, you recently expressed admiration for C.C. Myers and the way he and his construction company so quickly got the collapsed span of freeway in Oakland rebuilt after a tanker truck fire brought it down, and how he had the replacement connection span of the Bay Bridge with Yerba Buena Island ready to go and inserted it neatly over Labor Day weekend, in 2007.

Elsesser: I think that's the way of the future. He has proven it can be done. Why does the construction process take so long? Faster construction saves money. It saves cost due to disruption. Give the contractor a very big bonus for completing the project quickly. The Empire State Building was built in a short amount of time, fifteen months from groundbreaking to ribbon-cutting—they just did it. Our construction process has become too slow, too many things getting in the way of the achieving the essential goal of making something.

Reitherman: Often people point to the risks of hurrying up construction, because perhaps something will be skipped over in the rush. Is there also the possibility for quality improvement as well as getting it done faster? We all have known deadlines that have made us concentrate more intensely.

Kang: The priorities become clearer. It is easier to identify what is not really essential and avoid those diversions.

Elsesser: You can get everyone motivated to achieve an exciting goal, and of course, there is the monetary reward if bonuses are spread around.

This relates to some of our discussion about BSD and building systems. Those were very rapid projects to develop a system for organizing and constructing buildings, and then the buildings were built very rapidly. We seemed to be able to do things faster then. Now it takes five years to plan a project, and another five years to talk about the plan.

Structural Inspirations from Nature

Kang: You have done some presentations with all the structural examples drawn from things like leaves and coral, the architecture and structure of nature.

Elsesser: The microstructures of nature are fascinating, and more sophisticated than what we do in structural engineering.

Throughout history, buildings have been thought of as an extension of the ground, rather than following a flexible approach. The story above the ground is thought of as rigidly moving with the ground, the floor above that moving along with the one below it, and so on, but

12 Bruce Graham was the SOM architect who completed the design of the Inland Steel Building. Chuck Bassett, of the San Francisco SOM office, completed the design of the Zellerbach Building.

we know the reality is that there is great movement within the structure whether we like it or not. Engineers know that there will be those movements, but the design concepts don't recognize that basic fact.

Kang: You've talked about the integration of structure and architecture. What do you see happening in that area now?

Elsesser: A good example is the Deutsche Post Tower in Bonn, Germany by Helmut Jahn. Two crescents are offset, in ten-story increments. All the glazing on the sunny side can be opened, and it lets air up a ten-story chamber to solve the environmental problem. Groundwater is used to dissipate built-up heat and cool the building. In this country, there are very few clients who want to spend the money to do the best possible building. It takes an enlightened owner, and an enlightened design team.

Chapter 9

Engineering Education and Research

In teaching at Stanford, I started at about 2500 BC and worked my way up from there.

Kang: You have taught at both Berkeley and Stanford. How did that come about?

Teaching at U.C. Berkeley

Elsesser: I was talking with Claude Stoller one day, who was on the architecture faculty at U.C. Berkeley. I gave a lecture in his class, and then the following year taught a course on seismic design at Berkeley. My seismic design class was not how to do calculations. I asked the architecture students, "What do you want to accomplish? Let's see what structural solutions will make that possible."

Reitherman: Eric, you also served as the structural instructor in the graduate studio course, where each three-person team of architecture students would design a building and produce the architectural, structural, mechanical, and electrical working drawings. The student team I was on chose to do a retrofit project, using the original powerhouse on the campus, essentially a big unreinforced brick barn of a building. I would go through a whole pad of that green-tinted graph paper doing calculations of the out-of-plane and in-plane strength of the brick wall with various thicknesses of concrete applied and alternative reinforcement patterns. Eric would come in the next week, glance at the calculations, politely push the pad aside, and proceed to provide intelligent shortcuts for how to design an efficient system—and how to save a lot of paper. [Laughter.]

Elsesser: Yes, I did that working drawings course one year for the lead instructor, the architect Howard Friedman, who was the father of David in our office.

Arnold: Howard also taught the practice course, which was very popular with the students.

Engineering Research in Academia

Arnold: What do you think of the relationship between academia and practice, Eric? Does academia support the practitioners, or just academics?

Elsesser: There are some engineers with experience in practice who are great at doing research. Egor Popov was one. He was focused on the big design picture. Vitelmo Bertero is another. But there are many in academia who are so narrowly focused that their work is not as relevant to the practice of engineering. We don't need to refine mediocre structural solutions with more research. We need research on improvements.

Nick and I always made an effort to stay up with the research done at the universities. The key thing was to find out which of the systems had the best seismic performance, rather than only looking at what the code allowed.

We learned a lot from Egor Popov, from the shake table testing at by U.C. Berkeley at their Richmond Field Station near the campus. We learned things from the universities—Michigan, Illinois, and Berkeley in particular produced useful research. We thought a lot of the work of a young PhD from Berkeley, Helmut Krawinkler, who asked Nick and me for letters of recommendation to be on the faculty at Stanford, and we willingly did so. And as of now, 2007, Helmut has just retired.

Kang: What research did you find particularly interesting?

Elsesser: All of Popov's work was valuable. He did fundamental research that developed the seismically detailed steel moment-resisting frame. His research was later unfairly criticized for misleading the practicing engineer into thinking steel moment frames were indestructible, and in the 1994 Northridge earthquake there were fractures. But he did his testing on an early version of the steel moment frame, smaller sections and so on, and then the industry went on to extrapolate too much when details and layouts changed.

Kang: What about Popov and the eccentric braced frame? .

Elsesser: A marvelous idea. If Popov were still around it would be used more.

Teaching at Stanford

Kang: Did teaching allow you to lift your head up from the work on the drawing board in the office and look around?

Elsesser: Yes, it opens your mind. When I started to teach at Stanford, that came about a little differently. For the 1998-1999 school year, I became the Shimizu Visiting Professor at Stanford. It was up to me to propose something. So I did five major lectures, each with about 280 slides of buildings, projected in pairs. They weren't lectures on structural engineering. They were lectures on buildings. I talked to the students about how buildings are actually constructed. I bought about a hundred books and copied illustrations from them to illustrate the points.

In teaching at Stanford, I started at about 2500 BC and worked my way up from there. What are the building types? Why did it take more than a century to build a gothic cathedral? Sometimes there was a design concept for a cathedral but no one knew how to do it yet. I had a lecture on concrete construction, another on steel construction, one on bridges. Bridges are fascinating. I started with the simplest of bridges, like a single rope strung over a stream, and going on to very complex bridges designed by Santiago Calatrava. I finished with a lecture on contemporary structures. I would show photographs of sailboats and ask them if they could design a sailboat. Why not? It's a structure. If you understand the forces of the wind and the water, the reactions and balance of the structure, you should be able to understand its structure. I am giving that collection of slides to Stanford. Helmut Krawinkler will oversee that.

Reitherman: An Eric Elsesser slide show is instructive concerning engineering, and it is also a work of art. Same with every talk I've ever seen Chris give. And by the way, we commonly use the word "slide" now to refer to an electronic image projected from a laptop computer. PowerPoint slides are not the same quality as the photographic film transparencies. You were referring I presume to photographic slides, like Kodachrome?

Elsesser: Yes. I have my own photographic set-up for copying from drawings, and also enjoy taking pictures in the field. I worked at mixing up color and black and white in my lectures. A little text followed by ten photographs with no text, rather than each slide having some bullet points on it with a small photo. PowerPoint tends to lead to plodding through one similar image on the screen after another, with too many words.

History of Engineering

Arnold: Eric, you've always been interested in the history of engineering. In architecture, even today, history is very, very important. But in engineering, there is very little attention paid to history.

Elsesser: Unless you dig through books on your own, the engineer gets no sense of engineering history. It shouldn't be that way. Engineers need to understand how structural systems have evolved over time and why. The course I taught at Stanford was essentially the history of structural engineering as revealed through architectural history.

Kang: How did you balance the teaching with the work in the office?

Elsesser: You can do a lot in a 14-hour day.

Lectures

Arnold: You'll recall, Eric, that we did a series of lectures around the country for architects. They were initially funded by NSF, later by FEMA, and they were organized in different cities by the AIA, the American Institute of Architects. This was in the early 1980s. They were an introduction to the whole scope of seismic design.

Reitherman: How successful were those lectures in getting architectural converts, so to speak, to your approach of close collaboration between architect and engineer in seismic design?

Arnold: I'd have to say there was very little effect.

Elsesser: I would agree. The people who showed up at our seminars were often already interested in the subject, sometimes a principal, or a retired architect who was interested in the subject of earthquakes. But the designers we wanted to reach were usually back in the office designing.

Reitherman: Has the circle of architects interested in the subject of earthquakes grown over the years?

Elsesser: No. For most any engineering subtopic in earthquake engineering, you can come up with dozens of names of engineers interested in any topic, no matter how narrow. But it takes a big damaging earthquake to get any architects interested in earthquakes—then there are a few articles in the architecture journals. After a few months, the interest fades away.

Reitherman: One of the prominent Arnold-Elsesser lectures was at the 50th anniversary Annual Meeting of EERI, which had a historical theme to it about the development of different seismic design approaches and structural systems.

Elsesser: That was a lot of fun. We used a lot of illustrations for that.

Reitherman: Another prominent EERI lecture, Eric, is your current 2007 Distinguished Lecture, the annual honor and lecture tour of EERI. What have you been lecturing about, under the title of "What's Around the Corner in Seismic Design"?

Elsesser: I gave one in San Diego, St. Louis, the San Francisco area, trying to get people to think of new ways to do things in the future.

Reitherman: Another lecture and award, and one that ties back to your first job in structural engineering with John Blume as well as your education at Stanford, was the 2005/2006 Distinguished Lecture of the John A. Blume Earthquake Engineering Center. That one was entitled "Improving Seismic Safety and Performance of Buildings Through Innovative Structural Engineering."

Elsesser: That audience was mostly students. One of my lectures for fellow structural engineers that I recall is the one I gave on the 1985 Mexico City earthquake for the SEAOC Annual Convention held in Ixtapa, Mexico, 1992. That earthquake was very instructive.

Reitherman: How about owners, people who aren't architects or engineers?

Elsesser: After the Loma Prieta earthquake I lectured to the residents of Sausalito, and there was a big crowd. That was a matter of simply explaining to them what kinds of vulnerabilities houses in that town might have.

Reitherman: What about briefings to owners while projects were underway?

Elsesser: We've done lots of those. Also presentations that preceded the project and actually led to the project. For example, we gave several lectures on City Hall to the City of San Francisco people, because we had done enough analysis to know some of the problems of that building and how it would behave in the next earthquake.

Reitherman: What sort of questions do owners ask?

Elsesser: They don't ask anything. They don't really care, until you hit them with a verbal hammer and say something like large portions of their building may fall down on the 3,000 occupants.

Kang: Perhaps that sounds like a scare tactic, but it's really just delivering scary news.

Arnold: How do you persuade a building owner to do a responsible job on a retrofit or a new building?

Elsesser: In my experience, any owner who is going to keep the building for a long time is concerned about earthquakes. You talk to them about earthquakes and they listen. Anybody who is building a building in order to sell it is not interested. It's as simple as that.

Reitherman: What about lectures to a completely different kind of audience: did you ever give a talk to gradeschool kids?

Elsesser: I was invited to lecture to the class of my grandchild, Adam's son, when we were doing the San Francisco City Hall project. That was a field trip to the building, so they could see the actual project.

Kang: How did you explain base isolation to grade school kids?

Elsesser: I did a dance showing the different movement with and without the isolators.

Chapter 10

Thoughts on Architects and Engineers

There is a big difference between architects and engineers who are concerned only with design and those who are involved in the construction or are also the contractors.

What Makes a Good Architect?

Kang: You're interested in space, and art, and sculpture, and in my opinion that's one of the gifts of good architecture—it is art in space.

Elsesser: But there's too much routine work that goes on in architecture. I never regretted not being an architect, but I very much like architects. The project wouldn't exist without the architect.

Kang: What makes a good architect?

Elsesser: It takes somebody who really understands the brief of the project, the issues and requirements, takes the time to think about it, and then talks about it with consultants, before deciding which way to go. If the architect decides too much too soon—without sitting down with the structural engineer—you get problems.

There are so many bad buildings going up, like the new residential tower at the San Francisco end of the Bay Bridge. A horrible design.

Kang: Architectural or structural design?

Elsesser: Both. It is so near to the bridge that any cladding damage that results in falling debris is a serious threat. The developer wanted it that way, but it is the condominium owners who are going to be the risk-takers over the years.

Kang: In a high-seismic area, you feel you need a different architecture?

Elsesser: Absolutely. San Francisco is not Chicago.

Architects and Engineers

Kang: What about some of the architects and engineers whose work you have seen constructed in your lifetime? I know you have mentioned the projects of Ove Arup from time to time. Did you know Peter Rice?

Elsesser: He was one of their very good engineers, no question about it. But most of Ove Arup's work is not based on seismic design—either it's for a nonseismic region or for seismic regions the conception doesn't come together as well as I would like. It's a matter of really thinking of what will happen to the building in the real earthquake. It's not just a matter of running a computer analysis and taking the results for the end result. You have to think about what might happen that your analysis isn't including. Peter Rice did a beautiful job of designing train stations in places like England and France. When you don't have to worry about strong earthquakes, all kinds of options open up.

You have to consider what happens when the primary structural system starts to fail. You have to consider backup systems. That's what the owner really wants, even if they don't know it.

Another creative designer to mention is Santiago Calatrava, a delightful architect to listen to. I've heard him lecture a few times. He is very creative and inventive in his work. He combines architecture, engineering, and fine art. I don't know the details of the seismic aspect of his projects.

Kang: Who have been some of the engineers here in San Francisco whose work you have respected?

Elsesser: Steve Johnston at SOM I've already mentioned as a creative engineer. Nick and I thought he got a bad deal getting bounced out of the role of chief structural engineer in the San Francisco office of SOM. Navin Amin was another good engineer who the SOM firm didn't want to make a partner, so he quit, and is now with Middlebrook.

Ron Mayes has always been a good engineer, advocating isolation but with good technical reasons, and we've used him many times as a consultant.

Sig Freeman is a good engineer I have known from the first day he walked into John Blume's office, starting a few years after I was hired there. I still talk with Sig frequently.

Henry Degenkolb did great work. Chin and Hensolt also. Egor Popov I've mentioned. We loved working with him, and Vitelmo Bertero was a pleasure to work with also.

Jerry Weisbach is an architect I've known for many years. He later became an attorney, running one of the better legal firms here in town.

We did one job for Cesar Pelli's office, here in the Mission Bay development in San Francisco, the community center building. He has done very high quality work, but perhaps is less creative than someone like Fazlur Khan or Santiago Calatrava.

We worked with the I.M. Pei office in doing a Stanford project of five buildings. That worked out nicely. They wanted to put stone cladding on the building with a steel moment frame holding it up, but we convinced them that the structural and nonstructural system would be incompatible.

Mario Botta was a delight to work with, and the San Francisco Museum of Modern Art was a fun project. He adjusted to the structural guidance we were giving him and agreed to use steel framing for the structure of his building and to use precast brick-clad panels instead of laying up brick walls, from the ground up.

Kang: Talk about some former engineers and architects whose work in the twentieth century is also marked by structural expression.

Elsesser: Pier Luigi Nervi was an absolutely marvelous designer, and he built his own designs. That made the difference. He made models. He checked everything out. It was a great combination—design plus construction.

Eero Saarinen wasn't a constructor but he knew how to build, and his buildings show it. He did some experimentation in his work.

Félix Candela was a marvelous experimenter with concrete shells. The interesting thing is that there was a national ASCE conference here in San Francisco and everybody wanted to hear about his shells. He said he wasn't going to talk about shells, because they leaked. They did okay in the dry climate of Mexico, but it wasn't a formula to be repeated.

Robert Maillart, along with Nervi, was one of the great designers of concrete. His goal was to reduce the structure to the absolute minimum. And he had that concern with how the structure would be built, not just how to design it.

Eduardo Torroja was the same. Along with Nervi and Maillart he was one of the three greatest engineers we have seen in the realm of concrete.

Antonio Gaudi's buildings are unbelievable to see, and he used structural models to guide his designs.

Kang: You seem to point out that these engineers were all hands-on in their work. They either were construction contractors plus engineers, or worked hard on how to get their structures built. What about some of the prominent names in twentieth century architecture, like Frank Lloyd Wright, Le Corbusier, Walter Gropius, Alvar Aalto, Mies van der Rohe?

Elsesser: That same point applies to architecture. There is a big difference between architects and engineers who are concerned only with design and those who are involved in the construction or are also the contractors.

We all know about Frank Lloyd Wright. He always produced a special design for his building that could not quite be built the way he wanted. He couldn't quite pull it off, technically. Most of his buildings had to be repaired and strengthened.

Corbu was very, very good. He was a marvelous designer, along with Gropius. Aalto and Mies were also great architects.

So when you look at all these great architects, what is striking to me is that the buildings of Frank Lloyd Wright didn't really work. He had a well-deserved reputation for his design talent, but these other architects had real buildings that performed their roles and didn't have problems. And Wright had a civil engineering educational background. Perhaps he should have relied on consulting structural engineers who had stronger wills.

Architectural Styles

Kang: What about entire architectural styles and eras, perhaps in connection with engineering and construction? You dealt with broad patterns in your presentations at Stanford when you were a Visiting Professor there.

Elsesser: Let's take the Duomo in Florence, built in the Renaissance. Marvelous. An absolute gem, can't be beat. So creative that nobody could dream of it beforehand, and then they had to wait a hundred years to get somebody smart enough to figure the dome out and complete it.

The Gothic buildings were special. It usually took a couple hundred years to construct one of those big cathedrals. They were doing some very special structural things with their vaults, and that was, in essence, experimentation.

Further back, with Byzantine architecture, we have to talk about Santa Sophia in Istanbul, Turkey. The dome there is magnificent. They built it a little too flat and it partially collapsed. They raised the dome, and they needed to put in additional buttresses. It is one of the engineering marvels of the world to go see in person. You look up and see this vast dome floating, with all the little windows around the drum under it.

Before the Byzantine period, the Romanesque and Roman periods are interesting. I've been to Rome about a half dozen times, intentionally seeing different buildings, and also about half a dozen other times to other parts of Italy. Sylvia made it to Europe first, with her grandparents in the mid 1960s, when I was younger and was in Alaska roughing it, sleeping on the ground. She was staying in first class hotels of Europe and I was writing her letters about chipmunks scurrying across the floor.

Much later, Sylvia and I stayed in an apartment in London for a month, in 1969. So we went to Europe for a month. I always took the attitude that a vacation was vacation. You didn't take any work with you, didn't make any phone calls—you were on vacation. You work hard to get things in order, but then you just go. I found that a vacation kept refreshing for several months afterward.

Chapter 11

Collecting and Appreciating Art

The only reason I've had the time to work on my engineering, and also on my interest in art, is my wife, Sylvia.

Kang: The Forell/Elsesser office at Battery and Pine Streets in San Francisco currently has a lot of great art on the walls and pieces of sculpture, which sometimes surprises visitors.

Arnold: Yes, it is a great art collection.

Kang: And Eric is the curator. [Laughter.]

Reitherman: I remember the first time I walked into your former office on Clay Street in San Francisco during that BSD project on building configuration and seismic design funded by NSF,[13] when you were the structural advisor to the project. I thought, "Wow, this is like a really nice architect's office!" There were some beautiful drawings of bridges on the wall that an architect might have had on display.

Elsesser: You spend a lot of time in the office, you might as

13 Christopher Arnold and Robert Reitherman, *Building Configuration and Seismic Design.* John Wiley and Sons, New York, 1982.

well make it a nice place. I went to New York and while walking down the street found a shop with some old prints of bridges. I bought about ten and had them framed. Those are the ones you remember.

Sculpture

Kang: Before we talk a bit about your art collection, tell me a little bit about the structural engineering of art you have done, the structure of sculpture. How did that come about?

Elsesser: I've been the structural consultant to Aris Demetrios for about thirty years now. We [Forell/Elsesser] were the engineers for a building across from the Buena Vista Bar at Hyde and Beach. Demetrios was doing a sculptural entrance for the building and was referred to me to make it work structurally—and we hit it off. I've helped with a number of his sculptures. Currently, Mason Walters and I have been working on two 40-foot-high sculptures for the U.C. Merced campus.[14] The idea is that at graduations, one person at a time can walk between them.

Reitherman: So this is a lot more than anchor bolts you're talking about? You consult on the structure of the sculpture itself, to make it strong and stable, not just how to anchor an object?

Elsesser: Yes, that's what I mean—the whole metal object is a sculpture, but also a significant structure. Demetrios makes some large, unusually shaped objects. He did a 50-foot-tall sculpture by Interstate-80 east of Sacramento. Here's a model of one of his sculptures that was designed to be larger than this room. Watch how it vibrates. It moves back and forth this way, then it flexes the other way. This way, that way, alternating. The trick in that case was replicating that intended motion at a much different scale, an interesting problem in structural dynamics.

Demetrios has some very sound ideas. I just suggest some slight modifications to improve balance and so forth, and then do the detailed analysis of his complete design.

Kang: You've talked about some of the triggering events that have gotten you interested in design and construction, going back to your father's work as a cabinetmaker. What are the other influences on your interest in design?

Elsesser: Back at Stanford, in my senior year, I knew I wanted to take an art course. So I went over to the art professor, whom Sylvia and I later had design our wedding rings. He is still teaching and is in his 90s—Matt Kahn. I asked him to let me take his course, to make mobiles, and he said okay. We had weekly meetings. I had a book on Alexander Calder and was interested. By the end of the quarter, I had about twenty of these mobiles. I put a vice on the table of the dormitory room shared with three other guys, made it a workshop, and tried to get them to go to the library every night to let me use the room for my artwork. At the end of the quarter, all my large objects took up half the exhibit

14 The sculpture, *Beginnings*, was completed in June, 2009, and students walked through the sculpture at their graduation ceremony.

space allotted for the entire show of student coursework.

I like large sculptures, and in San Francisco, there aren't any great ones, only second-class ones. We've gone to see the big sculptures of Richard Serra in New York and Los Angeles—great stuff. I'm every bit as turned on by sculpture and art as I am by engineering.

Photography

Kang: You take wonderful photographs. How did your interest in photography start?

Elsesser: When I was twelve, I went on a trip around the Southwest with my family for two months with a little Brownie camera. I took lots and lots of photographs. None of them turned out because the camera wasn't working. [Laughter.] So I was forced to get a decent camera. In junior high and high school, I was using the darkroom. Later on, when I was married, we always turned one of the bathrooms into a darkroom.

Reitherman: Do you think you can learn discipline from film photography, wet photography, developing your negatives and prints in the darkroom, even if you now also have digital cameras at hand?

Elsesser: Yes, and there's also the quality of photographs, the range of values still being so much greater in film photography, for example. We've talked about how that affects the projected image, real photographic transparencies versus digital images in a PowerPoint show.

Book Collection

Kang: What about your book collection? You have some great books, and some rare ones?

Elsesser: I have given them all to Forell/Elsesser Engineers. I prefer books instead of online viewing on a computer.

Kang: What was it about the books? The images of architecture and engineering?

Elsesser: Yes, and it is reminiscent of when I was a student at Stanford, using the library.

Art in the Elsesser Residence

Elsesser: Sylvia and I over the past twenty years, as we've worked on and lived in our house here overlooking San Francisco Bay, have had the opportunity to make it a very interesting place for art, but we live in it, so it's not a museum.

Kang: What is it that appeals to you and Sylvia about particular pieces of art?

Elsesser: We're attracted to slightly different things, but basically, I would say 90 percent of the time we agreed on the things we liked, which is pretty good.

For example, this glass sculpture, based on an Egyptian canopic jar, is from Scottsdale, Arizona. Sylvia and I were at the opening of a show in a gallery. We were on opposite sides of the room. I pointed to the glass sculpture. She pointed to it. We put a hold on it overnight and came back the next day and bought it.

This wooden sculpture of a seated gentleman—it takes the form of a chair—is one we saw while we were walking in Venice. The shop was closed, but we tapped on the window and got the attention of the craftsman inside. He didn't want to open up, but with sign language—pulling out my wallet and waving lira notes at the piece—he understood we had a strong desire to buy that particular item and opened up. It was a nice chance event.

Kang: Can you characterize these various pieces of art you have collected in terms of what attracted you to them?

Elsesser: We like the craftsmanship. If we don't like the craftsmanship, we're not interested. We have a bowl from Scottsdale that has lines in it that the artist had started to use in her work that are just marvelous. We have several pieces by Ruth Asawa, who met Trude, my stepmother, at Black Mountain College. When Trude was here in California, she was the head of weaving at the California College of the Arts and Crafts in Oakland. Ruth used to knit, with wire, to form fabulous little structures that would gradually take shape. We would come over and visit and there would be these large spools of wire, and she would knit and talk. Some are now at the De Young Museum in San Francisco.

Kang: She did that fountain on the steps of the hotel just uphill from Union Square, with the sculpture of all the figures cast into the side of the fountain.

Elsesser: Ruth put a representation of my father's house at the top of that sculpture, on the south side. [See page 73.]

Kang: What about that sculpture here in your house that looks like an Egyptian sculpture of a cat?

Elsesser: This is one of the earliest pieces Sylvia got. The sculptor made a statue of a cat that was taller than the ancient original. When the sun comes through in the morning in the winter and hits it, only for a few days, it just glows. We always liked the spatial relationships of the works of art in our house. For example that bowl sat in its shipping case for about three years most of the time. I would take it out and bring it up to put it somewhere, but we didn't find the right place. Finally we put in the middle of a room, but when we got the sculpture of a horse, that filled that space. We never thought in advance of where we would put a particular object.

Kang: What strikes me is that so many of these art works are sculptural.

Elsesser: Absolutely, they have to be.

Kang: And the way they use materials?

Elsesser: Absolutely. The reason this house, with the view out over San Francisco Bay, is located this way, is that it was built by a retired navy captain who wanted to see the whole Bay. He had epaulettes, mustache, tall hat, the whole bit. Along with the house across the street, these were the first houses in this area.

Sylvia Elsesser: In the Prohibition era, he had a job spotting rum-runners delivering to various coves.

Elsesser: And Sylvia, I have to say that thanks to you, you made it possible for me to do everything that I could do.

Sylvia Elsesser: I knew you could do it. It takes courage, it takes talent. And long hours—you worked seven days a week, and I took care of the kids and household on my own.

Elsesser: The only reason I've had the time to work on my engineering, and also on my interest in art, is my wife, Sylvia. She's run a beautiful house. We've been married now for fifty-three years.

Kang: That's a long time. So you've been able to pursue your interests and know that you had an anchor at the house?

Elsesser: Yes.

Kang: Or perhaps also a rudder?

Elsesser: Yes, both. [Laughter.]

Photographs and Illustrations

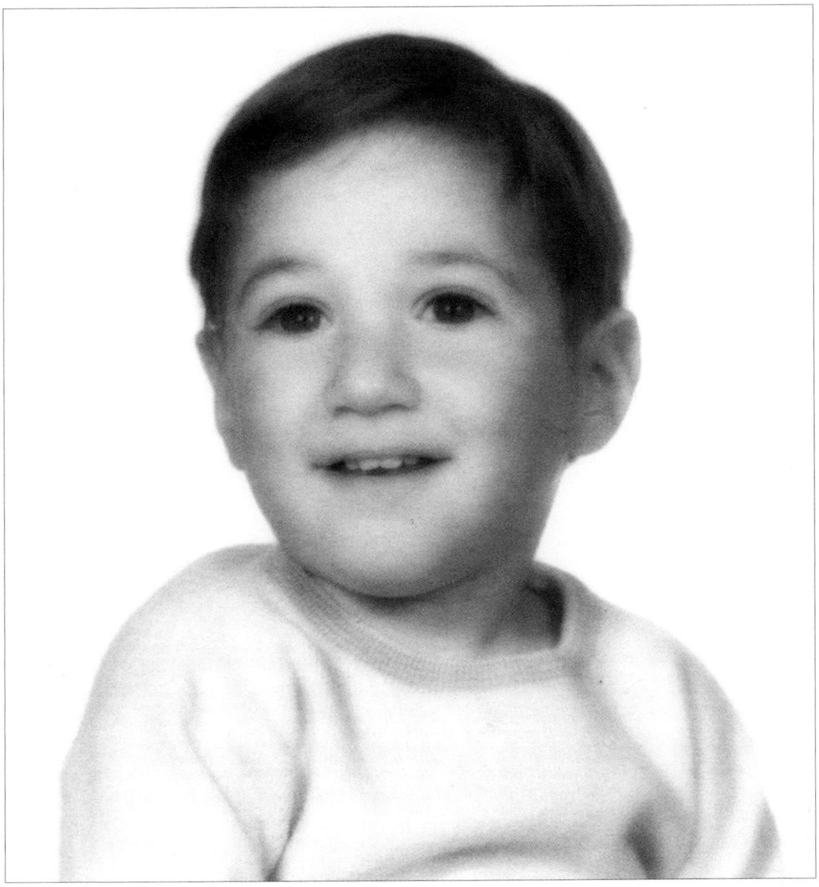

Eric Elsesser as a toddler.

John Elsesser, Eric's father, and Trude Guermonprez, Eric's stepmother.

Eric in grammar school.

Eric on his high school track team.

Eric in his Stanford University dormitory room with the mobile structures he made in art class during his senior year.

Sylvia and Eric at Eric's twentieth birthday party.

Sylvia Levin and Eric Elsesser at their undergraduate graduation from Stanford, while they were engaged. They were married a few months later, in September 1955.

Eric at work at the engineering firm of John Blume in San Francisco, circa 1956.

The house of Eric's father, 810 Clipper Street, San Francisco, incorporated into the fountain sculpture by Ruth Asawa at the Hyatt Hotel on Stockton Street, near Union Square, San Francisco.

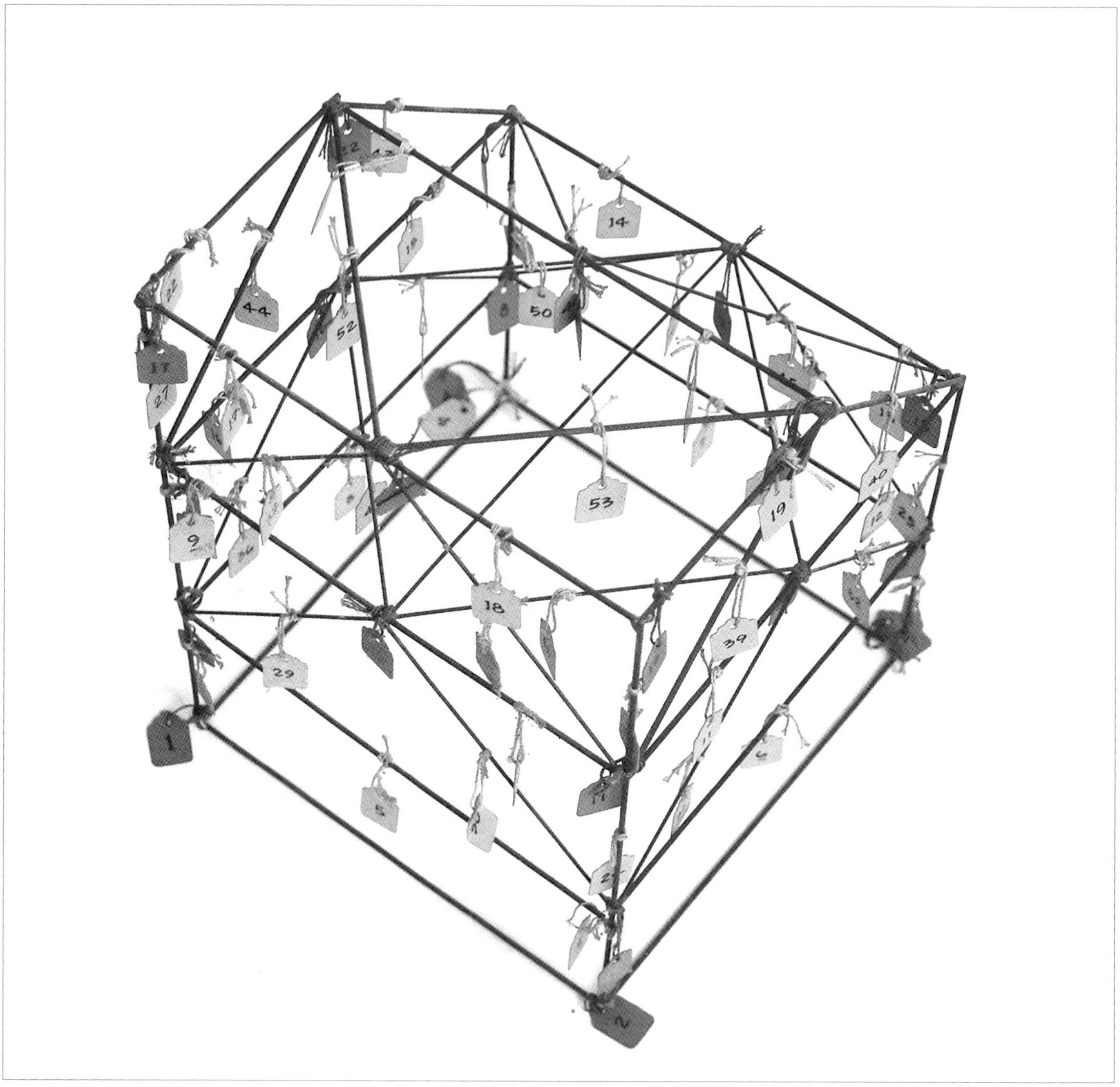

A wire-frame model by Eric Elsesser,
done before computers generated such images.

A concept for a spiral highrise structure, derived from a wooden sculpture.

A sea plant in Eric's office—a motivation for finding an organic framing form.

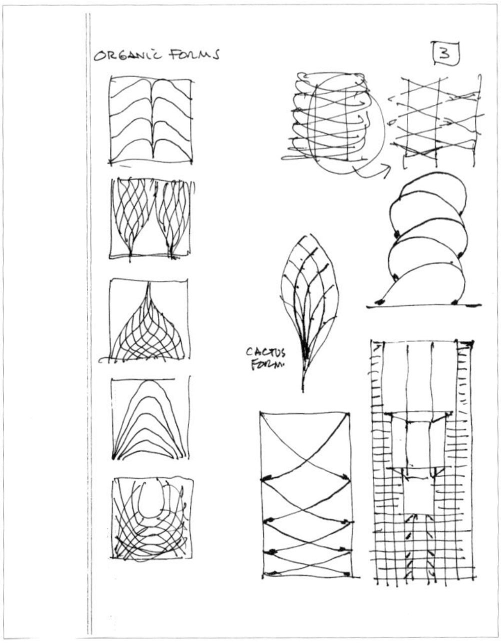

Sketches by Elsesser

Left: a development of ideas from organic forms to structural systems.

Below: natural structural forms found in trees and plants.

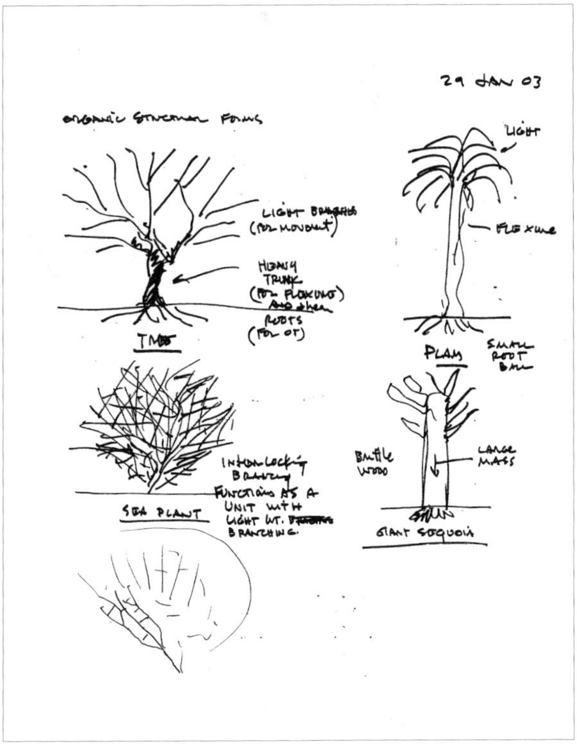

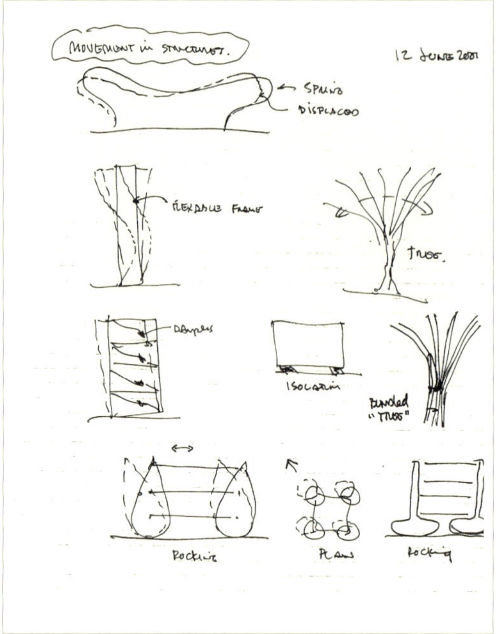

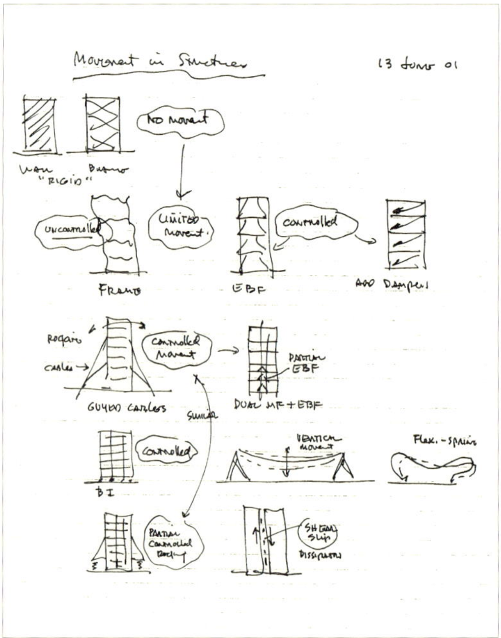

Sketches by Elsesser inspired by organic forms: structural concepts of controlled and uncontrolled movement. Eric often generated very quick sketches as he brainstormed.

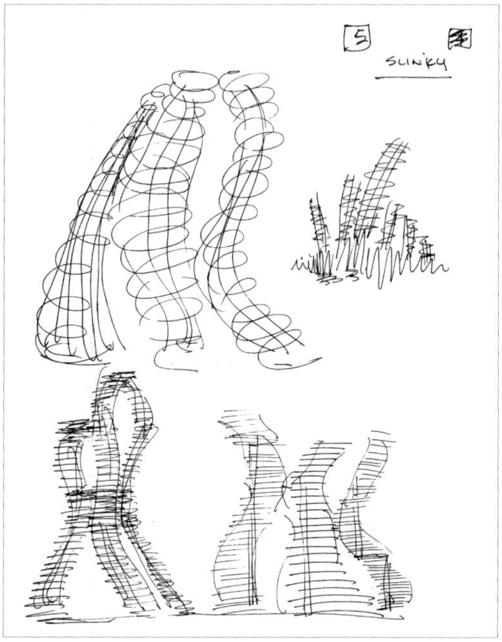

Sketches by Elsesser: structural forms inspired by the concept of a coil.

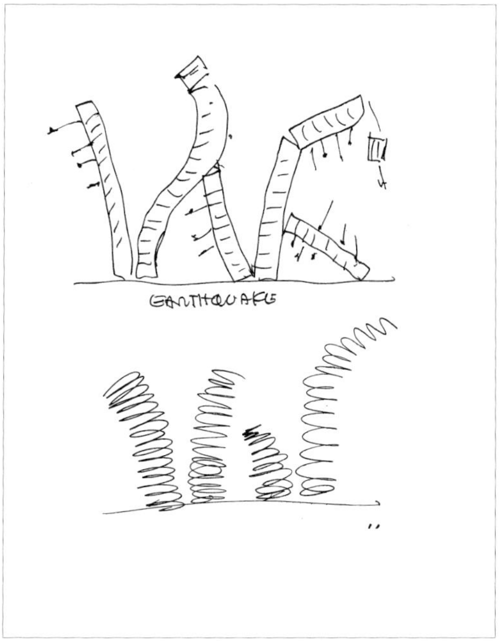

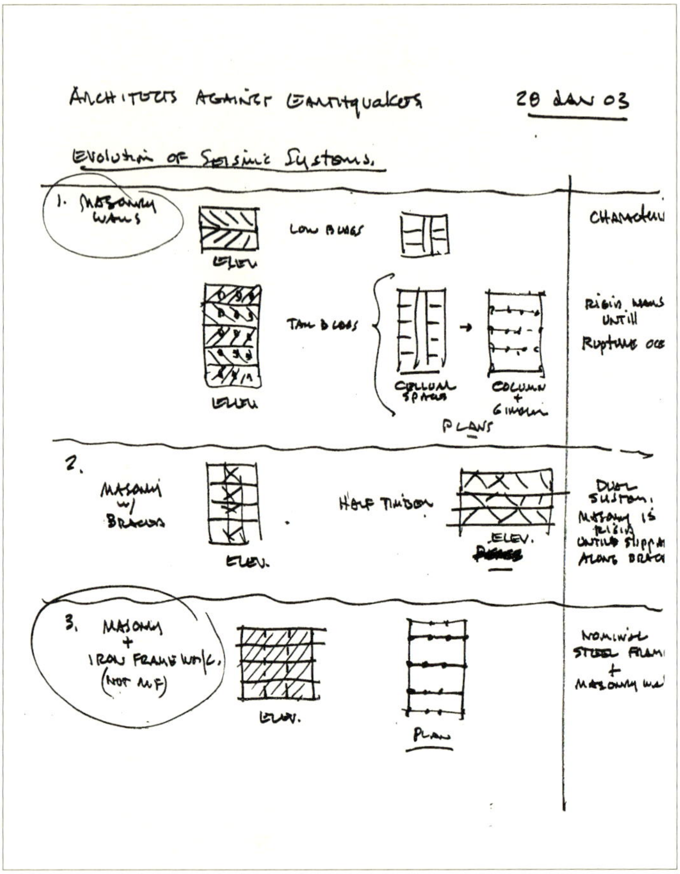

Sketch notes by Elsesser involving the evolution of seismic systems—type of system, elevations, plans, and notes.

An Elsesser sketch of the San Francisco City Hall, used to explain to the client how the building would "dance" on its isolators during an earthquake.

EARTHQUAKE RESISTING SYSTEMS

SELECTED PROJECTS (1960 – 1991)	BEARING WALL			DUAL				BRACES			MOMENT FRAME					OTHER	
	MASONRY SHEAR WALL	CONCRETE SHEAR WALL	TIMBER SHEAR WALL	FRAME & INFILL WALL	DUAL CONCRETE FRAME & SHEAR WALL	DUAL STEEL FRAME & SHEAR WALL	DUAL CONCRETE FRAME & MASONRY SHEAR WALL	STEEL BRACED FRAME	ECCENTRIC BRACED FRAME	DIAGONAL SPACE FRAME	CONCRETE DUCTILE MF	STEEL DUCTILE MF	CANTILEVER COLUMNS	CLUSTER MF	STEEL TRUSS MF	COUPLED CONCRETE SHEAR WALL	BASE ISOLATION
RODEF SHOLOM TEMPLE, SAN RAFAEL	●																
STANISLAUS COUNTY JAIL	●																
FORT ORD BARRACKS		●															
McALLISTER STREET HOUSING, S.F.		●															
CALIFORNIA STATE PRISON, VACAVILLE		●															
NORTHERN CALIFORNIA WOMEN'S DETENTION FACILITY		●															
COMMERCE CLEARING HOUSE, SAN RAFAEL		●															
ELMWOOD DETENTION FACILITY, SAN JOSE		●															
MARRIOTT HOTEL, SAN RAMON		●															
ST. FRANCIS SQUARE HOUSING			●														
OAKS COLLEGE, UCSC			●														
STERN HALL ADDITION, UCB			●														
UCLA PARKING STRUCTURE				●													
U.S. COAST GUARD BARRACKS, ALAMEDA				●													
BETHLEHEM TOWER HOUSING, SANTA ROSA					●												
FRENCH HOSPITAL MEDICAL OFFICE BLDG., S.F.					●												
FROMM & SICHEL BUILDING, S.F.					●												
CONTRA COSTA COUNTY DETENTION FACILITY						●											
IRVINE THEATER, UCI						●											
PSYCHOANALYTIC INSTITUTE, S.F.							●										
STATE OFFICE BUILDING, SANTA ROSA								●									
CABRILLO SCHOOL, S.F.								●									
MILL VALLEY SCHOOL								●									
CHEVRON ORTHO RESEARCH CENTER, RICHMOND								●									
THEO LACY DETENTION CENTER, ORANGE CO.								●									
SYNTEX CORP. COMPUTER CENTER								●									
FEDERAL EXPRESS HEADQUARTERS								●									
SAN JOSE FEDERAL OFFICE BUILDING									●								
UCLA MEDICAL OFFICE BUILDING									●								
CALIFORNIA STATE UNIVERSITY, LIBRARY II									●								
CHEVRON PARK, SAN RAMON									●								
SAN FRANCISCO STATE STUDENT UNION										●							
CMGM, STANFORD UNIVERSITY											●						
SAN FRANCISCO ART INSTITUTE											●						
CAL POLY ARCHITECTURAL SCHOOL											●						
SONOMA STATE STUDENT UNION											●						
FRENCH HOSPITAL											●						
IBM ST. TERESA PROGRAMMING CENTER, SAN JOSE												●					
IBM CORPORATION, ALMADEN												●					
PASADENA POLICE BUILDING												●					
ST. JOSEPH'S HOSPITAL, STOCKTON												●					
U.S. COAST GUARD TRAINING FACILITY												●					
UCLA AMBULATORY CARE COMPLEX, OB												●					
FILBERT LANDING OFFICE BUILDING, S.F.												●					
SANTA ROSA FEDERAL OFFICE BUILDING												●					
SAN MARIN HIGH SCHOOL													●				
BANK OF STOCKTON													●				
LAFAYETTE CHURCH													●				
TOMALES HIGH SCHOOL													●				
INTEL ASSEMBLY BUILDING, SAN DIEGO														●			
SANTA CRUZ GOVERNMENTAL CENTER														●			
PETALUMA GENERAL MAIL FACILITY														●			
SURGE LABORATORY, UCSF															●		
HEWLETT-PACKARD ASSEMBLY / LABORATORY															●		
LIFE SCIENCE BUILDING, UCB																●	
PHYSICAL SCIENCES BUILDING 2, UCI																●	
SALT LAKE CITY & COUNTY BUILDING																	●
OAKLAND CITY HALL																	●
U.S. COURT OF APPEALS & POST OFFICE																	●
SAN FRANCISCO CITY HALL																	●
SAN FRANCISCO PUBLIC LIBRARY																	●
SAN BERNARDINO COURTHOUSE																	●

The evolution of the seismic design philosophy of Forell/Elsesser Engineers office as illustrated by selected projects and their structural systems.

Principals and staff of Forell/Elsesser as of 2000 at San Francisco City Hall.
Front row, left to right: Simin Naaseh, Jim Guthrie, Eric Elsesser, Elizabeth Halton, Bill Honeck.
Second row, behind Naaseh's left shoulder, are Paul Rodler, David Friedman, Grace Kang.
Top row, third from left, is Mason Walters.

Eric Elsesser, 2001. Photo credit: Vera Photography.

Index

A

Aalto, Alvar 57
Albers, Anni 3
Albers, Joseph 3
Amin, Navin 56
Analysis, relationship to design 10, 25
Architects 55
Architectural Resources Group 36
Arnold, Christopher 19, 34, 52
Art and artistic influences 2–3, 13, 59–63
Arup, Ove 46, 56
Asawa, Ruth 62, 73

B

Base isolation 33–41, 53
Bassett, Chuck 47
Beckett, Welton 46
Benjamin, Jack R. 6–7
Bertero, Vitelmo V. 44, 45, 50, 57
Black Mountain College, North Carolina 3
Blume, John A. 9–11, 18, 21, 46, 52, 56
Blume, John A. Earthquake Engineering Research Center, Stanford University 52
Bolotowsky, Ilya 3
Botta, Mario 27, 57
Bowles, Chester 17
Brunnier, Henry J. 21
Buckling-restrained braced frame (BRBF) 28–29
Buildings designed by Forell/Elsesser Engineers. *See* Forell/Elsesser Engineers, major projects
Building Systems Development (BSD) 19, 41

C

Calatrava, Santiago 51, 56, 57
Calder, Alexander 60
Campbell and Wong 14
Candela, Felix 57
Canfield, Robert 37
Chin and Hensolt 57
Configuration of buildings xv–xvi, 10, 44–45

D

Degenkolb Engineers 21
Degenkolb, Henry J. 57

de Kooning, Willem 3
Demetrios, Aris 60
Dynamic Isolation Systems (DIS) 35

E

Earthquakes
  Loma Prieta, California (1989) 26, 34, 36, 41, 52
  Mexico City, Mexico earthquake (1985) 52
  Northridge, California (1994) 24, 29, 50
  San Fernando, California (1971) 24
Eccentric braced frames 28, 40, 50. *See also* Structural systems
Ehrenkrantz, Ezra 19, 34
Elsesser, Adam (son) 2, 11
Elsesser, Eric
  art and artistic influences 2–3, 13, 59–63
  Blume, John A., first job with 9–11
  book collection 61
  education
    early 1–2
    Stanford University B.S., M.S. 5–7
  engineering, approach to 18
  establishing own firm 13–15
  family 1–3, 11
  Forell/Elsesser Engineers. *See* Forell/Elsesser Engineers
  lectures and presentations 47–48, 51–53
  photographs of 65, 67–72, 83-84
  photography as a hobby 61
  sculpture 60–63
  sketches by 77–81
  Stanford University B.S., M.S. 5–6
  Stanford University, teaching at 51
  U.C. Berkeley, teaching at 49
Elsesser, John (father) 1, 66, 73
Elsesser, Linnea (daughter) 11
Elsesser, Rikee Richterman (mother) 2
Elsesser, Sylvia L. (wife) 2, 6, 9, 11, 25, 58, 60–63, 70, 71
Elsesser, Trude Guermonprez (stepmother) 2–3, 62, 66
Engineering
  conceptual versus analytical approach 3, 18
  history of 51
  practice and academic research 50

F

Forell/Elsesser Engineers
  early days of partnership 17–21
  major projects 23–31, 33–40, 43–48
    Bethlehem Towers, Santa Rosa, California 25
    Davies Symphony Hall, San Francisco, California 25
    Ford assembly plant, Richmond, California 27
    French Hospital, San Francisco, California 24
    IBM, Cottle Road, San Jose, California 25
    Salt Lake City and County Building 34–35
    San Francisco City Hall 35–36, 39, 53, 81, 83
    San Francisco Museum of Modern Art 27, 57
    Student Union, San Francisco State University 25
  Principals 20–21, 83
  seismic evaluations 25–26
Forell, Nicholas F. 15, 17–21, 26, 44, 50, 56
Framing systems, comparison of 28–31. *See also* Structural systems

Freeman, Sigmund 56
Friedman, David A. 20–21, 83
Friedman, Howard 50
Fuller, Buckminster 3

G

Gaudi, Antonio 57
Gropius, Walter 3, 57
Guermonprez, Trude (stepmother) 2–3, 62, 66
Guthrie, James 20, 83

H

Halton, Elizabeth 20, 83
Healy Tibbetts 14
Hellmuth, Obata, and Kassabaum (HOK) 27
Hirsch, Ephraim 18
History of engineering 51
Honeck, William 20, 83

I

Irons, Tony 38
Irregular buildings 10, 44–45

J

Jacobsen, Arne 27
Jahn, Helmut 48
Japan, earthquake engineering in 34
Johnston, Steve 10, 56
Jokerst, Mark 20

K

Kahn, Fazlur 57
Kahn, Matt 60
Kang, Grace 20, 83
Keatinge-Clay, Paffard 23, 24
Klee, Paul 3
Kline, Franz 3
Koo, Rosalyn 20
Krawinkler, Helmut 50, 51

L

Le Corbusier 23, 57
Lectures and presentations 47–48, 51–53
Levin, Sylvia. *See* Elsesser, Sylvia L.
Loma Prieta, California earthquake (1989) 26, 34, 36, 41, 52

M

Maillart, Robert 57
Marquis and Stoller 13–15
Marquis, Robert 13–15
Mayes, Ronald L. 35, 36, 56
McCue, Boone, Tomsick 19
McCue, Gerry 20, 44
Mexico City, Mexico earthquake (1985) 52
Modeling of structures 1, 10, 46–47, 74
Moment frames 28–31, 44, 50, 57
Motherwell, Robert 3
Muto, Kiyoshi 30
Myers, C.C. 47

N

Naaseh, Simin 20, 83
Nervi, Pier Luigi 57
Netsch, Walter 47
Neutra, Dion 24
Neutra, Richard 24
Nonstructural elements 26, 28, 34, 44, 57
Northridge, California earthquake (1994) 24, 29, 50

P

Park, Robert 41
Paulay, Thomas 30, 41
Pei, I.M. 57
Pelli, Cesar 57
Performance of structures 29, 40, 45, 52
Popov, Egor P. 28, 40, 45, 50, 57

R

Reaveley, Lawrence D. 35
Reid, John Lyon 20
Research and engineering practice 50
Rice, Peter 56
Richterman, Rikee (mother) 2
Rockwell and Banwell 20
Rodler, Paul 20, 37, 83

S

Saarinen, Eero 57
San Fernando, California earthquake (1971) 24
San Francisco City Hall, San Francisco, California 35–36, 39, 53, 81, 83
San Francisco Museum of Modern Art 27, 57
Scott, Stanley vii–viii
SEAOC Blue Book 10–11
Seismic design
  computer programs 45
  modeling structures 1, 10, 46–47, 74
  philosophy of 43–48, 82
Seismic evaluations 25–26
Seismic isolation. *See* Structural systems, base isolation
Serra, Richard 61
Severini, Gino 3
Sharpe, Roland 10
Sketches by Elsesser 77–81
Skidmore, Owings & Merrill (SOM) 10, 25, 46, 56
Stanford University
  B.S., M.S. from 5–6
  teaching at 51
Steel. *See* Structural systems, steel
Stoller, Claude 13–15, 49
Structural Engineers Association of California (SEAOC) Seismology Committee 10–11
Structural systems 24
  base isolation 33–41, 53
  buckling-restrained braced frames, 28–29
  comparison of 28–31
  constructing quickly 47
  dual 28, 39–41
  eccentric braced frames 26, 28, 40, 50
  evolution of 28–31
  irregular 10, 45
  modeling 1, 10, 46–48, 74
  moment frames 28–31, 44, 50, 57
  nonstructural elements 26, 28, 34, 44, 57
  performance of 29, 40, 45, 52
  steel frames 24, 27, 28–31, 34, 38, 44, 46, 50, 57

T

Tarics, Alex 20, 34
Thiel, Charles C. Jr. xiv
Torroja, Eduardo 57

U

University of California at Berkeley, teaching at 49

V

van der Rohe, Mies 57

W

Walters, Mason 20, 35, 46, 60, 83
Weisbach, Jerry 57
Wildenhain, Marguerite 3
William Clayton Building, Wellington, New Zealand 34
Willsea, Fred 18
Wong, Worley 14
Wright, Frank Lloyd 57–58